جنود مثاليون

PERFECT SOLDIERS

جنود مثاليون

PERFECT SOLDIERS

The Hijackers: Who They Were,
Why They Did It

Terry McDermott

HarperCollins*Publishers*

HarperCollins books may be purchased for educational, business, or sales promotional use. For information, please write: Special Markets Department, HarperCollins Publishers Inc., 10 East 53rd Street, New York, NY 10022.

FIRST EDITION

Designed by Lundquist Design, New York

Printed on acid-free paper

Library of Congress Cataloging-in-Publication Data is available upon request.

ISBN 0-06-058469-6

05 06 07 08 09 ❖/RRD 10 9 8 7 6 5 4 3 2 1

For Mac and Betty

HE WAS THE PERFECT SOLDIER: he went where you sent him, stayed where you put him, and had no idea of his own to keep him from doing exactly what you told him.

—Dashiell Hammett, *The Dain Curse*

Contents

Key Figures

The Hamburg Group

Mohamed el-Amir aka Mohamed Atta: September 11 pilot, leader with
Ramzi bin al-Shibh, of Hamburg group; Egyptian

Marwan al-Shehhi: September 11 pilot; Emirati

Ziad Jarrah: September 11 pilot; Lebanese

Ramzi bin al-Shibh, aka Omar: leader with Atta of Hamburg group, tried
to become a pilot, coordinator of September 11 attacks; Yemeni

Said Bahaji: member of Hamburg group, fled before September 11;
German-Moroccan

Zakariya Essabar: member of the Hamburg group, tried to become a
pilot, fled before September 11; Moroccan

Mounir el-Motassadeq: member of the Hamburg group, accused of
assisting the September 11 plot; Moroccan

Mohammed Fazazi: imam and mentor to the Hamburg group; Moroccan

Mohammed bin Naser Belfas: mentor to Hamburg group; Yemeni-
Indonesian

Mohammed Haydar Zammar: Al Qaeda possible recruiter of Hamburg
hijackers; Syrian-German

Mamoun Darkazanli: Zammar associate, radical Islamist, mentor to
Hamburg group; Syrian-German

Abdelghani Mzoudi: friend of Hamburg group, accused and acquitted of
assisting September 11 hijackers; Moroccan

Abdullachman al-Makhadi: mentor of Ziad Jarrah, imam of Greifswald
Mosque, friend of Belfas and Zammar; Yemeni

Mohammed Ragih: member of Hamburg group; Yemeni

Bashir Musleh: friend of Jarrah in Greifswald and Hamburg; Jordanian

Abbas Tahir: friend of Musleh and Jarrah; Sudanese

Friends and Acquaintances

Shadi Abdallah: friend of Hamburg group, member of al Tawhid,
    Jordanian terror group; Jordanian
Shahid Nickels: friend of Hamburg group, drifted away;
    German-South African
Ahmed Maklat: friend of Hamburg group, left Hamburg out of fear;
    Sudanese
Yassir Boughlal: college classmate and friend of Zakariya Essabar,
    resisted recruitment; Moroccan
Aysel Sengün: Ziad Jarrah's girlfriend; German-Turk

Al Qaeda

Osama bin Ladin: leader of Al Qaeda; Saudi
Khalid Sheikh Mohammed: mastermind of the September 11 attacks;
    operational commander of Al Qaeda; Pakistani
Muhammed Atef: Osama bin Ladin's lieutenant; Egyptian
Ayman al-Zawahiri: Bin Ladin lieutenant; Egyptian
Abu Zubaydah, né Mohammed Hussein Zein-al-Abideen: Bin Ladin
    lieutenant captured in March 2002; Jordanian
Tawfiq bin Attash, aka Khallad: Bin Laden lieutenant, coordinator of
    USS *Cole* bombings; Saudi
Hambali, aka Encep Nurjaman, Riduan Isamuddin: leader of Jemaah
    Islamiyah, a Southeast Asian terror organization and Al Qaeda
    coordinator in the region; Indonesian
Yazid Sufaat: Hambali deputy, assisted Moussaoui, Hazmi, Mihdhar, and
    Khallad; Malaysian
Ali Abdul Aziz Ali: Khalid Sheikh Mohammed nephew, funneled money
    to 9/11 hijackers from United Arab Emirates; Pakistani
Mustafa al-Hawasawi: financial facilitator of September 11; worked with
    Ali Abdulaziz Ali in UAE; Saudi
Imad Eddin Barakat Yarkas, aka Abu Dahdah: leader of Madrid cell;
    Syrian
Zacarias Moussaoui: accused 9/11 conspirator, awaiting trial in Virginia
    on suspicion of wanting to become a hijacker; French-Moroccan

Manila Air Bombing Campaign
Khalid Sheikh Mohammed: mastermind of the September 11 attacks;
Pakistani-Kuwaiti
Ramzi Yousef, aka Abdul Basit Abdul Karim: organizer of the 1993
bombing of the World Trade Center; planner of Bojinka, the
Manila airline bombing campaign; Pakistani
Mohammed Jamal Khalifa: Bin Laden brother-in-law, alleged terror
financier in Philippines and elsewhere; Saudi
Abdul Hakim Murad: participant in Bojinka, the Manila airline bombing
campaign; Pakistani
Wali Khan Amin Shah: participant in Bojinka, Manila airline bombing
campaign; Afghan

Afghanistan
Abdullah Azzam: Palestinian leader of the Afghan Arabs; Palestinian
Gulbuddin Hekmatyar: fundamentalist warlord; Afghan
Abdur Rasul Sayyaf: fundamentalist warlord; Afghan
Zahed Sheikh Mohammed: Khalid Sheikh Mohammed's brother, head of
Kuwaiti charity Lajnat al-Dawa Islamia; Pakistani-Kuwaiti
Abed Sheikh Mohammed: Khalid Sheikh Mohammed's brother, killed in
Battle of Jalalabad; Pakistani-Kuwaiti
Aref Sheikh Mohammed: Khalid Sheikh Mohammed's brother;
Pakistani-Kuwaiti
Burhanuddin Rabbani: fundamentalist, ethnically Tajik political leader;
Afghan
Mullah Omar: Taliban leader; Commander of the Faithful; Afghan

September 11, 2001
American Airlines Flight 11, Attacked the North Tower of the World Trade Center
Mohamed Atta: pilot, Egyptian
Abdul Aziz al-Omari: Saudi
Satam al-Suqami: Saudi
Wail al-Shehri: Saudi
Waleed al-Shehri: Saudi

United Airlines Flight 175, Attacked the South Tower of the World Trade Center
Marwan al-Shehhi: pilot, Emirati
Ahmed al-Ghamdi: Saudi
Hamza al-Ghamdi: Saudi
Fayez Banihammad: Emirati
Mohand al-Shehri: Saudi

American Airlines Flight 77, Attacked the Pentagon
Hani Hanjour: pilot, Saudi
Majed Moqed: Saudi
Salim al-Hazmi: Saudi
Nawaf al-Hazmi: Saudi
Khalid al-Mihdhar: Saudi

United Airlines Flight 93, Intended to Attack the Capitol, Crashed in Pennsylvania
Ziad Jarrah: pilot, Lebanese
Ahmed al-Nami: Saudi
Ahmad Ibrahim al-Haznawi: Saudi
Saeed al-Ghamdi: Saudi

Preface

IN NOVEMBER 2001, on a blustery winter day in Hamburg in northern
Germany, a young woman, the wife of one of Mohamed el-Amir Atta's
old roommates, talked about an image she couldn't get out of her head.
When the American war against Afghanistan had started that autumn,
when the bombs began falling and people began dying by the score, she
would sit in front of her television, staring in disbelief, unable to compre-
hend that the conflict in a very real sense had been set in motion by her
husband's old roommate, Mohamed.

Watching the explosions, she would try to match them, the war,
everything that had gone on in the world since September 11, to her memory of the slight young man padding around his student apartment in his
shower shoes. It didn't fit. Mohamed was a tough guy to figure and she
never liked him, but this, all of this because of Mohamed? It's impossible,
she told herself. Not little Mohamed in his blue flip-flops.

There is much about Atta, one of the September 11 hijackers, and his
brethren we can't now know. But when a person moves through the
world, he leaves a path that can be traced, however faint parts of it may
grow. In the Atta traces, the image that lingers is of a man who was far too
small to accomplish the huge thing he did. There is something deeply
unsatisfying about this. We want our monsters to be outsized, monstrous.
We expect them to be somehow equal to their crimes. More than anything,
we want them to be extraordinary, to allow us to believe the horrible thing
they did is unlikely to be repeated. In its own odd way, this is a comfort-
ing thought. When we go looking for people capable of inflicting such
great harm, the last thing we expect to find is little Mohamed in his blue
flip-flops.

This is, foremost, a reported book about the men who executed the September 11 attacks against the United States. The aim of the reporting was to discover and attempt to understand those men and the places, people, and ideas that shaped them. Not unusually for a large news event, a narrative of the attack and attackers was constructed with astonishing speed: by the end of the first week after the attacks, the central story had been set and the characters cast. The September 11 attackers were caricatured either as evil geniuses or as wild-eyed fanatics. Unfortunately, as is also usual in big news events, much of the initial information was either factually wrong or, more commonly, irrelevant and misconstrued. While there might well be elements of both of these extremes in some of the men, they were largely neither of these things. The intent of this book is to try to come to a better understanding of who these people were, and thereby understand why they did what they did.

Most of the men of September 11 came from apolitical and unexceptional backgrounds. They evolved into devout, pious young men who, over time, drew deeper and deeper into Islam. As they did, they debated endlessly how best to serve their God, how to fulfill what they came to regard as sacred obligations. They saw themselves as soldiers of God, which prompts the obvious question: What kind of religious belief could empower men to inflict such great harm and deprivation on other men, women, and children? The inquiry that grew from it yields a truly troubling answer: the men of September 11 were, regrettably, I think, fairly ordinary men. I say this is regrettable because it was their ordinariness that makes it much more likely there are a great many more men just like them. In the end, then, this is a story about the power of belief to remake ordinary men; it is a story about the dangerous power of ideas wrongly wielded.

I'm certain some will take this as an attack on Islam. It is not. When I was young, I was fashionably enamored of the idea that the world was a very sordid place and could use some fixing up, according, of course, to my own architecture. I believed I knew what needed to be done. This wasn't a religious belief in the least; it was utterly secular. In fact, I suspected religion was a significant part of the problem. Then, early enough in my career to matter, I went on assignment to Cambodia. This was after the wars and the Khmer Rouge had come and gone, leaving nothing but ruin and bleached bones behind. It was without close contest the saddest place I had ever been; and it had become that way as the result of profound secular belief, of someone's certainty of ownership of the one true way. It is this certainty, not the belief itself, that causes the problems.

I'm sure this effort to understand the motivations of the men behind September 11 will upset other people who will think that any attempt at understanding is somehow an attempt to excuse, or even glorify. It is not. A primary task—and great joy, too—of the journalist is to empathize, to try to understand the way the world appeared to the people being written about. I have tried to do that here. The attacks of September 11, 2001, were horrific events, world-altering, life-changing, and, for far too many, life-ending. As I write now, three years later, the killing continues with no good end in sight. The sooner we come to understand what is happening, the sooner we will have a chance to stop it. Until we do understand, we have no chance at all.

جنود مثاليون

PERFECT SOLDIERS

Welcome

E SPECIALLY IN WINTER, AL QUDS when unkind winds blow hard and wet down from the neighboring northern seas, Hamburg can be a nest of dark prospects. On a typical harsh gray day, when the city grows dim and the evening turns cold, when the men leave the bleak streets and come into Al Quds Mosque, they bring their own warmth with them. Hands are clasped; bearded cheeks brush bearded cheeks; shoulders are clutched firmly, and the murmur of soft words and low easy laughter fills the vestibule.

Al Quds occupies a warren of sparsely decorated rooms above a body-building parlor in a poorer quarter of Hamburg, Germany's richest city, on Steindamm, a tough, seamy street that runs east from Hauptbahnhof, the city's central rail station. Steindamm offers its own addition to the winter chill in the form of the icy lipstick smiles of the Albanian hookers who lounge in doorways and mix with the dope dealers on the sidewalks below. The location is perfect for Al Quds: rent is cheap, the train station makes Al Quds accessible from all points on the Hamburg map, and Steindamm's more temporal pleasures remind the faithful of the decadence from which they are about to enjoy temporary reprieve. The men come to evening prayer from across the city and, more broadly, from across the Arab world. They are united often by language and always by belief and, here, as in many other places around the globe, by a piercing feeling—maybe not that far from heartbreak—of being on the outside of everywhere, looking in.

No social movement, revolutionary ideal, or earthly kingdom has surpassed in breadth and speed the triumphant spread of Islam from its seventh-century origin in the remote Arabian desert. In little more than one hundred years, the prophet Mohammed's word had reached westward across the rim of Africa, south to the equatorial center of the continent and as far north as the Pyrenees. Its dominion arced west from the Bosporus through the heart of Asia, blanketing much in between. Eventually, it washed over the great far eastern archipelagos, tying the Atlantic and Pacific Oceans with a broad band of Muslim lands.

Islam's authority in many of these places, reinforced by the Arab military conquests that often accompanied it, was absolute, dictating not just a system of belief, but a prescription for life in its every pitiful particular.

As the Arab empire of the Middle Ages receded and eventually disappeared altogether—even, or especially, in the Arab lands themselves—Islam remained where the empire's tide had carried it. For more than a millennium, Muslim lands were divided, subdivided, reassembled, and put asunder by conquerors from all directions, but no matter what the agent or angle of attack few of these lands ever fell from the ranks beneath the Prophet's banner. Muslim lands largely remained Muslim no matter who ruled their civil affairs and they remained so even as the Arab world sank into a period of long and, as became far too clear later, dangerous decline.

So it is that when Muslims—and in particular, Arab men—travel, many seek Islam's camaraderie and comfort wherever they go. When Mohamed el-Amir Atta, 24 years old and on his own for the first time in what had been a sheltered life, arrived in Hamburg from Egypt in the summer of 1992, one of the first things he asked for was the location of the nearest mosque. Amir was born and raised in an ambitious, not overtly religious middle-class household in Egypt, and by the time he arrived in Germany, had already earned a degree in architecture from the prestigious Cairo University. He had come to Germany for graduate school and, like many young Arab men abroad, was drawn to a more fervent and active embrace of Islam. He found his way to Al Quds.

There is within Islam, as they say, only one God and God is great, but any religion that requires its faithful to pray five times a day can expect them to exercise discretion in determining where and with whom those prayers are said. Metropolitan Hamburg has a sizable Muslim population, about 5 percent of its more than 4 million people, and mosques are spread throughout the city to serve it. The mosques, like churches in Christendom, segregate themselves by ethnicity, economics, and scriptural interpretation. The biggest, oldest mosques in Hamburg are Persian in origin and Shia in doctrine. The overwhelming majority of the rest are small, neighborhood mosques—often little more than storefronts; some, in fact, are actually in stores—that serve largely Turkish congregations. Al Quds is one of the city's few Arab mosques. Its version of Sunni Islam is harsh, uncompromisingly fundamentalist, and resoundingly militant.

The place itself is of modest size; its main prayer room holds at most 150 people. The walls are off-white, with Qur'anic verses painted on them in green. The carpets are gray, and the rooms have a utilitarian feel. There is a kitchen where men sit on metal folding chairs at flimsy tables taking simple meals of skewered meats, vegetables, tea, and soft drinks. Until the preaching starts, it has the convivial feel of a schoolboys' clubhouse, more sociable than revolutionary. But any innocence is burned away in the fierce torrents of words that have become the signature of the place.

In many respects, fundamentalist Islam is structured much like fundamentalist Christianity, which is to say there is very little real structure to it at all, little hierarchy and no absolute arbiters or authority. Guidance is contained in the core texts and the interpretations of these texts by historical figures. But the modern world is a complicated place, profoundly different than the world in which the founding figures of the religion lived. Things they could never have anticipated—recombinant DNA, rhythm and blues, liberal democracy—cry out for interpretation and the holy texts don't always present obvious assistance. Into this structural void every willing preacher walks with impunity, a virtual free agent, able to think and say whatever he believes. The preachers compete with one another for congregants. They rise and fall in popularity, and in the end answer to no one but the marketplace. Much of the influence of particular preachers is acquired through the circulation of video and audio tapes of their sermons. The most ferocious tend to be the most popular. They're in high demand and travel a well-developed European circuit.

Al Quds had been founded by Moroccans and regularly featured preachers from there. One of them, Mohammed Fazazi, was well known at home for his associations with outlaw terror groups. Fazazi, an imposing presence at 6-foot-3, equated the West's economic involvement in the Arab world to an occupation. He routinely advocated killing nonbelievers and implored his followers to embrace martyrdom, "to love death as much as the impious love life."[1]

Al Quds was distinctive in Hamburg but no different from thousands of other mosques around the world—from Riyadh to Yogyakarta to London. Throughout the 1990s, a new radical Islam was being promulgated in these mosques. It was a fire that had been building for decades, fed by the Islamic revolution in Iran and fanned by the wars that came later in Afghanistan, Chechnya, and Bosnia. This new Islam was based in equal parts on reverence for Qur'anic text and deep resentment at the place of Islam in the contemporary world.

The shabby mosque on Steindamm became a destination for young fundamentalist Arabs throughout Germany. Some, in fact, chose which university to attend based on proximity to the mosque. Even within its hard-core congregation there were divisions. At Friday prayer, when the congregation swelled and sometimes overflowed the main room, the hardest of the hard-core sat by themselves in a rear corner of the main prayer room, on the right as you entered. Over time, Amir came to join them. The group was small and fluid and mean. They threatened men who disagreed with them, beat those who wouldn't conform. Some men fled in fear; others were drawn to the flame and hardened by it. The small group produced three men—Amir, Marwan Al-Shehhi, and Ziad Jarrah—who, on September 11, 2001, led the devastating attacks against the United States and two other men who would have joined in the attacks if they had been able.

These men came from different backgrounds and countries, but in ways were strikingly similar. Most were from the fringes of whatever society they came from, whatever schools they had attended. All but one enrolled in German colleges and for many this was an imperfect fit. The men came to Germany at different times and to different cities over five years but were bound to one another by the ideas that painted the plain rooms of Al Quds. It was largely within these rooms that they met one another and their lives' last callings. It was there they heard preachers like Fazazi, who in one sermon told the congregants they must be prepared to kill everyone in their path. God demands it, he said. "Who participates in the war against Islam with ideas or thoughts or a song or a television show to befoul Islam is an infidel on war footing, that shall be killed, no matter if it's a man, a woman, or a child."[2]

Fazazi took an interest in the young men and met regularly with them after prayers to urge them along his path.[3] In these talks and in his sermons, he celebrated the new facts of life and death: "The jihad for God's cause is hard for the infidels, because our religion has ordered us to cut their throats and that we kill their heirs is a hard thing. . . . God the merciful has created the hell for the infidels as he created the paradise for the believers, too."[4]

The men would leave these sessions and the talk would continue among them for hours, for days, and eventually years. They would parse the slim distinctions they heard between this preacher or that, because it wasn't a question of a single man corrupting others. Not Fazazi, not Osama Bin Laden. There were many preachers saying some variation of the same dark things, many more than you could count. There was one thing they all agreed upon. This was war, they said, and they'd come looking for soldiers.

BOOK ONE
Soldiers

CHAPTER 1

A House of Learning

THE DELTA

NEARLY ALL OF EGYPT'S 65 MILLION people are squeezed by the great surrounding deserts onto thin ribbons of arable land strung along the length of the Nile River. This savannah, made fertile by the regular flooding of the river, has been populated for tens of thousands of years—far beyond the range of human memory. North of present-day Cairo, the river splits into two main branches—the Rosetta and Damietta—and innumerable smaller ones, a spiderweb of streams crisscrossing between the two larger channels. From there north, 100 miles to the sea, the river feeds a broad, improbably lush delta. These northern reaches of the Nile endowed one of the great civilizations of the earth long before the powerful realms of the western world were even the faintest of far-off dreams, when, as one Islamic scholar put it, "northern Europeans were still sitting in trees."[1] The Delta's abundance has forever remained the source of the enormous wealth and talent Egyptian civilizations have produced. Presidents, poets, and revolutionaries have all been shaped in its villages.

Today, the Delta remains Egypt's breadbasket. Its markets overflow; the roads are jammed with pickup trucks and donkey carts. Tractors are rare—most of the work of the fields is still performed the way it has always been, by hand and hoof. The Delta is thick with people, too. Women wear veils or scarves; many men wear the long cotton tunics called *galabiyas*, muddied at the hem from hard work on wet ground. The last village is seldom out of sight before the next slides into view. Between towns, the fields, small and irregularly shaped, jigsaw across the tableland. Billboards for the latest Nokia cell phones straddle irrigation ditches teeming with trash. Women bathe and wash dishes along the dirty shores.

Mohamed Mohamed el-Amir Awad el-Sayed Atta was born here in 1968 in the northernmost delta province of Kafr el-Sheik. His father, Mohamed el-Amir Awad el-Sayed Atta, came from a tiny hinterland village, and his mother, Bouthayna Mohamed Mustapha Sheraqi, from the outskirts of the provincial capital, also called Kafr el-Sheik. As was, and is still, customary in rural Egypt, the elder Mohamed and Bouthayna met and married by arrangement of their families. At the time of the wedding, Mohamed el-Amir, as he was known, was already an established local lawyer, having taken degrees in both civil and sharia, or Islamic, law. Bouthayna was only 14, but as the daughter of a wealthy farming and trading family, she came from several rungs up the social ladder and was a good catch for the ambitious Mohamed. They soon had two daughters, Azza and Mona, then a son named for the father.

They hadn't many relatives on the father's side and maintained a cool distance from Bouthayna's family. This was according to Amir's wishes, Bouthayna's family said. The father was regarded by his in-laws as an odd man—austere, strict, and private. He was and remains a bluff, forceful fellow who permitted little disagreement.

Village life in the Arab world offers much the same degree of privacy as village life elsewhere, which is to say, very little at all. Egypt's crowded geography further insists that life be communal and shared. People are piled on top of one another. To resist the weight of the centuries in which life has been spent and shaped this way takes real effort. Amir, a stubborn man, was willing to expend it.

"The father is alone. There are no brothers, one sister maybe. We never met her," said Hamida Fateh, Bouthayna's sister. "Here, the families are all very close. But even here, the father was separate."[2]

Fateh's family is prominent in Kafr el-Sheik; they own farmland, an auto-parts store, and a six-story commercial building. The family lives unostentatiously above a cobbled, dusty street in a cramped walk-up with whitewashed walls, plain rugs, overstuffed furniture, a Panasonic boom box, and a 19-inch Toshiba television. It is unair-conditioned and the apartment's balcony doors hang open to let the inevitable afternoon heat escape.

Fateh wears a head scarf, more out of habit than belief, she said; neither her family nor the Amirs were particularly religious. They were part of the secular generation that grew up in Gamal Abdel Nasser's Egypt, when the country's future did not seem as bound to the past as it does today. They were the generation that would remake Egypt and reclaim its glories. We are educated people, Fateh said, people from the country but not country people. Fateh studied agricultural engineering at university; her husband studied electrical engineering.

The senior Amir was ambitious, too, and exceptionally focused. His law practice thrived in Kafr el-Sheik, but he was not satisfied. "He moved to Cairo," Fateh said. "He wanted to be famous."

CAIRO

Cairo is slow to awaken most mornings. Dawn breaks on empty streets and the first layer of dust isn't raised for hours after. Breakfast is a rumor and it is common to find shops not yet open at noon. Cairenes see this as proof of a cosmopolitan sophistication, but in truth the city has not left the countryside that far behind. More than most great cities, Cairo remains today a collection of villages. A handful of those villages glitter in a ring around the metropolis; they're filled with jewelry shops, fashion boutiques, and Mercedes sedans. They are in constant communication beyond the surrounding deserts to the wider world. In this Cairo, fashion shows are staged on the green-treed lawns of European embassies, smoky bars serve chilled Russian vodkas, and lunch is taken on the terraces of century-old social clubs. Homes are guarded by iron gates and old men with automatic rifles. There is a circle of local society that dwells exclusively amid these shows, shops, clubs, and cafés, but most of the city does not even know these places exist.

The overwhelming majority of Cairenes spend their lives in that unknowing condition, tethered to their poverty, their past, and their God, who is, here as throughout Arab lands, ubiquitous, his presence announced through tinny loudspeakers five times a day, when the faithful and faithless alike are called to prayer. The prayer calls squeal through the cheap speakers and rattle down worn stone alleyways, bouncing off walls coated with grime a millennium in the making. In the teeming shanty-towns and walk-up apartment blocks, there are few guards and privacy is guaranteed only by implicit pacts among neighbors not to notice every-thing you do, say, and probably think, or in any event, to act as if they don't, because not noticing is next to impossible.

The gap between the two realms—that of modern, secular, ambi-tious Cairo and the vast, cramped, poorer quarters of the big city—often stretches into an abyss so broad people are seldom able to see, much less move, across it. On an island between two cultures—one ancient and suf-fused with myth and tradition, the other cool and sleek and as rational as a burglar—is the city's slight and struggling middle class. It was onto this island in Cairo society that the Amirs settled when they arrived from Kafr el-Sheik.

The family came to Eldmalsha Street in Abdin, a once grand, now faded, quarter near the old financial and government centers. By the time the Amirs arrived in the 1970s, the wealth of the city had already begun to migrate to newer districts, west across the Nile to Mohandiseen and Dokki and south and east to the newer suburbs of Maadi and Heliopolis. Old core neighborhoods like Abdin were left to crumble. Most of its five- and six-story stone apartment buildings were leftovers from British colonial rule, which ended, finally, in 1952. Lobbies were paved with rich marbles and limestones, but the tiles were chipped and broken with shards swept up into small piles in the corners.

Mohamed was 10 years old when the family arrived here. His father took advantage of the neighborhood's decline and leased a huge double flat that occupied an entire floor. This allowed all three children their own rooms, a rarity. The interiors of the old apartment were dim and still, the windows covered against the sun. Later, Mohamed's father bought a small vacation home up north on the Mediterranean coast, but the family lived frugally in town. The children's mother, Bouthayna, did her own cooking and cleaning. The father drove a used Opel, then traded up to a modest

Fiat. When Bouthayna's family came from the Delta to visit Abdin, they found the father had instilled his ambition in the children. "They respected their father's determination and demands on them," Fateh said. "It was a house of study. No playing, no entertainment. Just study."

The children weren't allowed to play outside the apartment. Young Mohamed's room looked out the back of the building, over rooftops and into a tangle of wires and adjacent windows. Neighbors said he used his window for clandestine conversations with neighbor boys. That was playtime. On the rare occasions they were allowed to watch television, said a cousin, Essam Omar Rashad, Mohamed would leave the room whenever belly dancing programs—staples of Egyptian broadcasting—came on.

"Mohamed's friends would sit on the corner there, chewing pistachios, spitting out the shells. Not Mohamed. There was no hanging around, no friends, very strict rules," said Mohamed Gamel Khamees, a neighbor who runs an auto repair shop on the ground floor of the Amirs' old building. "They came from a village, and they had their own traditions. They brought them along. They lived a closed family life. They were very polite but had little contact with any others."

One neighbor said the walk to elementary school, a mere 100 meters away, had been timed, and if the children took longer than the allotted few minutes to get home, they would be called to account. Another neighbor said they sometimes heard the father shouting at the children. "No one ever shouted back," he said.

Bouthayna had a little handcart she used when she went marketing. Neighbors laughed when they saw it. They thought she was putting on airs. It didn't matter, of course, what the neighbors thought. The Amirs knew what they were about. The family went its own way. The father was a husky, gruff man as likely to give a speech as an answer to a question. He was unapologetic about his lack of sociability. "We are people who keep to ourselves," he said. "We don't mix a lot with people, and we are all successful."

Abdin is one of the densest districts in one of the most densely populated cities on Earth. Life spills outside. The street becomes a place for entertaining, for sport, for business. When visitors come, chairs and a tiny foot-high table are brought out. Tea is served. Mohamed Khamees's repair shop—an arm's length from the tea table—is about the size of a walk-in closet. Work is done in the middle of the road. Down the block on the sidewalk is an auto body repair shop. There is no interior to it whatsoever. A lean young man with hands so soiled they look like black rubber gloves hand-sands dried putty to eliminate the crease in an old Russian Lada's hood. A donkey cart loaded with dates rolls by. A sweet potato salesman pushes his wagon past. In between the tea being poured and the sugar offered, a man rolls a whetstone by. The cries of the knife man, the date man and the sweet potato seller ricochet down the stone alleys. The body man pounds a fender and his hammer echoes in the long narrow gaps between buildings.

It is difficult to remain closed off here, even harder than in the Delta. Asked if Mohamed's family ever made exceptions—if, for example, it shared evening breakfast with neighbors during the holy month of Ramadan, which in Cairo is a period of daytime fasting and late-night socializing and celebration—Khamees said, No, the father was a tough man, not given to making exceptions. He insisted things be arranged his way, down to the smallest details. The family, Khamees said, was "like a set of rings interlocked with one another. They didn't visit and weren't visited." He paused for a moment and waved a hand at the insects circling the tea table's tiny sugar bowl. He looked up at the apartment. "Not even the flies entered there," he said. "Not even the flies."

UNIVERSITY

The Amir children were superior students. The girls, Azza and Mona, entered the science faculties at Cairo University, one of the most prestigious educational institutions in the Middle East. They continued their ascent there: Azza became a cardiologist, and Mona a professor of zoology. The university, based in the Cairo suburb of Giza, between the city and the desert, is mammoth, with 155,000 students and more than 7,000 teachers. It sprawls across both banks of the Nile, including an island in between. The campus is so large some students drive cars from class to class. Admission is granted solely on the basis of national tests. Degree programs are typically five years. The first year is a preparatory year, used to direct students into major areas of study. If you want to study medicine, for example, but your first-year grades are insufficient, you might find yourself—without consultation or consent—enrolled in the Department of Ornamental Horticulture.

Young Mohamed trailed his sisters through school to university, pushed along by their achievements and his dogged father. In the first year, students were assigned classes solely on the basis of their names.

"I found him standing there, staring up at the name sheets to see where he was assigned," said Mohamed Mokhtar el-Rafei. "I introduced myself. 'I'm Mohamed,' I said. So was he. We looked at the class sheets. We had three full classes of Mohameds. Oh, wow. . . . We used our fathers' names to refer to one another. I was Rafei. He was always Amir."

The two became friends. Both excelled in the first year, 1985, and were chosen for engineering, one of the most venerable and prestigious departments. It's hard to overemphasize the respect accorded engineers in much of the Middle East. People use the word as Westerners do "doctor," a title that becomes part of the name. The engineering department was immense; it had nearly 1,000 teachers. The size meant tremendous competition and—except for the very best students—little attention from professors. Within the engineering department, the highest-scoring students were assigned to the architecture program. The two Mohameds, whether or not they wanted to, would be trained as architects. Having achieved the great honor of admission to the august engineering department, for the first time in his life, Amir did not excel. Architecture, more than most creative disciplines, is a blend of the utterly pragmatic—what type of glass do you specify to keep heat out and let light in?—and the artistic—in what vocabulary should a house speak? Amir shone at the analytical subjects, but the curriculum at that time was skewed toward design and the more creative aspects of the discipline.

"He was a very clever person in mathematics, physical structures, less good in design and the more artistic aspects," Rafei said. "He had been one of the top-ranked students in high school, and he had a very high rank in his preparatory year. In our time, though, design was emphasized, and maybe you could say he couldn't adjust himself to what was needed. In the third year, when we studied soils, street plans, and steel, something more concrete, he excelled. . . . You would recognize him more as an engineer than an architect."

Another classmate recalled that Amir became upset when things didn't go his way. "He was a child," she said. "So like a child that one time something happened, where he didn't get the grade he wanted, and he pouted. Somebody said to him, 'You're acting like a child.' Then he got very, very angry, proving the point. He really was like a child. Spoiled."

Mainly, Amir is remembered as utterly normal. "Mohamed was there, sharing all our fun times. He liked it. He would tell jokes, laugh. He was one of us," said Waleed Khairy, a classmate.

The apparent calm of Amir's passage through childhood and university was at odds with the extraordinary tumult occurring within the country.[3] He and his friends were serious, bright students who were excited, in the way idealistic young architects often are, about the prospects of their profession, about the idea of building a new world. At the same time, thousands of other young Egyptians were serious about the idea of changing the world with somewhat less respectable methods.

Cairo is a plotter's paradise. There has long been a perpetual shortage of employment and a surplus of cafés where young jobless men join generations of others, idling away their days over sweet Turkish coffees, water pipes, and filtered Cleopatras, filling the cluttered streets with talk and a soft, smoky haze. Not all the talk is idle chatter.

Following independence from Britain in 1952, Egypt had enjoyed a period of great optimism and ambition. Gamal Abdel Nasser, the young Army officer who seized control of the country, was not content merely to break the bounds of colonialism; he wanted to break the longer economic and cultural decline the country and the whole Arab world had endured. He aggressively pursued modernization at home, reinvigorating the economy and beginning the long-haul task of lifting the millions of peasants out of poverty. He helped found the nonaligned movement, a coalition of largely third world nations that declared independence from the cold war rivalry between the United States and the Soviet Union and elevated Egypt's stature internationally. He played the U.S. and the USSR off one another, taking what assistance and charity he could connive from both. He reestablished Cairo as the intellectual center of the Arab world and, through uninhibited promotion of pan-Arab nationalism, raised hopes for political progress in almost all Arab nations. All things seemed possible. Too many things, perhaps.

In May 1967, the armed forces of Egypt, Jordan, and Syria massed along their borders with Israel. Nasser and others declared the time had come to wage a war to put an end to the Jewish state. War came, but with an utterly different effect. In less than a week, Israel drove Syrian forces out of the Golan Heights, Jordanians out of the West Bank and Egyptians out of the Sinai Desert and occupied all those territories. The war was an unmitigated disaster for the Arabs. Their horrible miscalculation of Israel's military prowess sent the Arab world spinning into a crisis of confidence from which, almost four decades later, it has yet to recover. At home, the loss provided crucial ammunition to regime opponents in what would become a permanent war between Islamist fundamentalists—that is, those who wished to remake their states as Islamic republics following strict sharia law—and secular modernizers. The secularists, it would be argued in venues scattered across half the globe for decades to come, had so weakened the will of their nations as to sacrifice their right to rule.

The problem in Egypt was even more acute than in the rest of the Arab world, in part because of Egypt's inheritance, the glory of its past. Unlike many of the countries of the Middle East, Egypt is not the arbitrary result of some colonial cartographer's whim. Its history is real—long and almost impossibly grand. But the glory of Egypt is far behind it. Few doubt that. When people wonder why, they find the easiest answer lies east across the Sinai Desert—Israel and, by implication, the United States, its sponsor. This broadly shared belief fueled the Islamist surge. Nasser by force of will and intellect had made Egypt the center of the Arab world; the failure of his military made it the center of fundamentalist reaction. His death and the rise of Anwar Sadat did nothing to ease this tension. Sadat was seen as an accommodationist puppet of the West. His grand achievements, the Camp David accords and the subsequent 1979 peace treaty with Israel, became exhibit A in the fundamentalist indictment against him. He retaliated with punitive crackdowns, jailing thousands, while at the same time saying he endorsed the Islamist cause, stating that Egyptian law would be based on Islamic law. Three years after Sadat's great triumph at Camp David, he was murdered by Islamists within his own army. They were so emboldened they did it not stealthily under cover of night, but in the most public place possible—at a parade.

Hosni Mubarak succeeded Sadat and continued the same schizophrenic approach to the fundamentalists: embrace at times; imprisonment, or worse, at others. Mubarak himself has been the object of more than three dozen assassination plots. To a significant extent, the repression and the radicalism created and reinforced one another. It was no longer possible to know which came first.[4] One clear result has been a political system that was democratic in name but harshly autocratic in practice. More than half the officially recognized political parties have at one time or another been barred from political activity. Strictly speaking, the involvement of religious groups in politics is entirely forbidden. Members of the strongest, most broadly active Islamist group—the Muslim Brotherhood—were routinely jailed for violating this prohibition. The lack of any avenue for legal dissent forced political opposition to the margins, almost ensuring that it would become extreme. The Muslim Brotherhood conducted major recruiting campaigns at Cairo University during Amir's college years. The Brothers called for a return to basic Islamic principles, including full implementation of Islamic law, and warned against the corrupting forces of modernization. The critique singled out Egypt's tilt toward the bourgeois decadence of the United States. The Brotherhood's campus activism coincided with a period of increasing religiosity in Egypt generally, which was part of the great Islamic awakening throughout the Arab world following the Iranian revolution.

The Amir household made conscious efforts to avoid the entire debate. The father, a devout Muslim, warned his children away from political Islam. Far from advocating a resistance to the West, he insisted that his son, in addition to his regular class work, study English, which he did at the American University of Cairo. The father clearly identified with the secular ambitions of the government. He stayed aloof from politics and public displays of religious belief and instead sought status and respect. Egytian lawyers, unlike their American counterparts, seldom got rich. Almost no one, in fact, got rich in Egypt without connections to Mubarak's ruling clique. Amir spent most of his career working for Egypt

Airlines and earned a decent, though far from spectacular, income. He joined one of Cairo's many social clubs, which since the days of British rule have been essential roadmaps of Cairene status. The country's ruling elite has taken lunch on the green terraces of the Gezira Club since colonial days. The only thing that changed after liberation, one diplomat said, was that Egyptian generals replaced British colonels at the club bar. Amir's club membership was emblematic. He couldn't climb onto the elite rung of the local ladder occupied by Gezira, the country's oldest and most prestigious club, but did manage to buy membership in the Shooting Club, one of the best of the aspiring second-tier, middle-class clubs.

Amir said he wanted his son to match his daughters' successes. He often drove Mohamed to and from university. This was not entirely unusual. Unless they move to another city, most young Egyptian adults remain in their parents' homes until they marry. Many 30-year-olds eat dinner at their mothers' tables every night.

Mohamed was graduated in the middle of his class, a lower rank than those achieved by his sisters and one that did not gain him acceptance into Cairo University's elite graduate school. His father gave him his used 1974 Fiat 128 coup as a graduation present, but insisted that he continue his studies. He told Mohamed that to be a success he needed to earn the title "doctor" in front of his name, as both of his sisters already had done. He urged young Mohamed to look abroad for graduate schools. Mohamed took temporary jobs as on-site construction supervisor on two building projects in Cairo and, at his father's insistence, enrolled in German language classes at the Goethe Institute.[5]

"My son is a very sensitive man; he is soft and was extremely attached to his mother. I almost tricked him to go to Germany to continue his education. Otherwise, he never wanted to leave Egypt," Amir said. "He didn't want to go. By pure coincidence, a friend of mine had visitors from Germany, two high school teachers in Hamburg. I invited them to dinner, and Mohamed was the king of the evening because he spoke German fluently."

Two weeks later, young Mohamed was in Hamburg asking directions to the nearest mosque.

CHAPTER 2

Alone, Abroad

HAMBURG

IN THE SUMMER OF 1992, Amir arrived at the north Hamburg home of the two middle-aged teachers his father had met in Cairo. He lived rent-free in an extra room in their small cottage. The couple had been organizing exchange programs between Germany and Egypt for years and were happy to help the bright young architect. He arrived with a single suitcase, but in other respects, he carried more baggage than almost anyone his hosts had ever met.[1]

Amir's family had been, by Egyptian standards, secular. But Egyptian standards did not obtain in Germany. Amir, even had he maintained his family's beliefs and religious commitment and simply transported them to northern Europe, would have been markedly more religious than the great majority of his German peers. Instead, as happens with many young people when they go abroad, his beliefs and the practice of them intensified. He strictly observed the requirement to pray five times a day, at a mosque if possible. He began observing a strict halal diet—no pork, no alcohol. Amir refrained from the pleasures young students often sought. He seldom socialized, never went to clubs or sporting events, and had little tolerance for any display of female sexuality. Even sleeveless blouses caused him to grimace.

Hamburg is a notably unrepressed city. Sex businesses—theaters, clubs, prostitution, publishing houses—thrive. For someone like Amir who would leave the room when belly dancers came on television, the rough and wild side of Hamburg was a rude introduction to the western world.

Hamburg's public licentiousness was a reflection of local attitudes that there were more important things to worry about. That belief was hard-won. Hamburg has been a city of remarkable resilience. Plague and cholera epidemics ravaged it in the Middle Ages. The commercial center burned to the ground in the Great Fire of 1842, then was carpet-bombed nearly to oblivion during a three-day Allied raid in the summer of 1943. After each disaster, it has been rebuilt, and is now Germany's wealthiest city and home base for its media elite. The only visible sign of its past difficulties are plaques affixed to buildings throughout the city stating the dates of their destruction and resurrection.

The Elbe River, its tributaries and the trade they carried were the city's original reasons for being, and its path to enrichment as the largest port of the medieval trading league of cities known as the Hanse. The Elbe cuts through contemporary Hamburg on a diagonal, connecting the city to the North Sea, 50 miles northwest, and in the opposite direction to Prague and the central European interior. The thousand-year-old harbor, with its great merchant ships, container cranes, and marshaling yards, remains central to the city's prosperity, but, like many working ports, exists in what might as well be an alternate universe for most of the 4 million people who live in the metropolis. If they see the harbor at all, it is in fleeting glimpses from the windows of passing trains and automobiles. The heart of the city, for most people, lies north of the harbor, interlaced with a web of canals connecting two central-city lakes to the river. The canals are lined with restaurants, hotels, clubs, offices, and boutiques. Hamburgers bustle among them, tall, fair, and stylish in cool Armani blacks and Jil Sander grays. Like the residents of many northern cities, the locals have a reputation for wool-wrapped aloofness. They move briskly to get where they're going and don't stop much to chat.

The port, as well, was the city's original connection to the Arab world and Central Asia. The trade remains substantial. For most of the last century, Germany's relationship to Arab lands has been the best among the European powers. This in large part was because Germany, alone among those powers, was not a colonizer of Arabia. The German and Ottoman empires abutted and competed along their flanks in southern Europe, but over time settled into a largely peaceful relationship, while France and Britain fought to wrest control of land from the Ottomans. Germany, enemy of the Arab world's enemies, became its friend. Relations continued through World War II, when general sentiment among Arabs favored a German victory. The war was regarded as a European affair, and Arabs sided with the one European power that had not colonized them.

After the war, Germany reemerged as an economic power. Among those who helped sustain that strength were millions of foreign workers, including for the first time in German history a substantial Muslim population. The majority were Turks or Kurds, but there were also Iranians and Palestinians, then later refugees from Yugoslavia and Albania. Islam became the third largest religion in the country.

In addition to the powerful magnet of its economy, Germany offered foreigners relatively easy and inexpensive access to its extensive university system. Technical education, engineering in particular, had become a conventional path of economic advance in Arab countries. German universities eagerly recruited foreign students.

Amir had come to Germany on a tourist visa. He would need a student visa to attend graduate school but apparently hadn't understood that he had to get it in Cairo; it could not be issued in Hamburg. The teachers, after Amir had moved in with them, went back to Cairo to make arrangements for other students they were helping come to Germany. While in Cairo, as a favor and without telling Amir, they put through an application for his student visa. When they returned to Hamburg and told him they had taken care of it for him, he grew quite upset.

"I am grown up now; I can take care of that myself," he told them.

"He said that a lot," his host said. "'I am abroad now; I am grown up. Now I can decide on my own.'"

It seemed silly to resist their help, "but that's the way he was," the woman said.[2]

Amir's hosts had traveled often to Egypt; at first, they celebrated Amir's cultural differences. The woman admired his seriousness. He was *eisern,* she said—iron. They often discussed religion. The woman knew the Old Testament well and tried to make the point to Amir that the roots of Islam and Christianity were intertwined; that in a sense they worshiped the same God. Mohamed would listen, then reply, yes, but what is written in the Qur'an is the truth, and, more importantly, the only truth. They would argue, the woman said, until, to avoid an outburst of her own anger, she would leave the room disgusted by his closed-mindedness.

Amir made few friends. He could be amiable and polite, but was never warm. His landlady felt that there was "always a wall between him and the family." She said that eventually she was made to feel uncomfortable in her own home. Amir complained when the landlady's adult, unmarried daughter came to visit and brought along her young daughter. It was strange, the landlady said. He played with the little girl and obviously enjoyed it. "He was free. The only time I remember him to be free," she said. But then he railed against the mother's lack of a husband and the presumed moral laxity that had produced the child. As he had in Cairo, if an objectionable program came on the television, Amir would leave the room in a huff. In the spring of 1993, by mutual agreement, Amir moved out of the little cottage.

HARBURG

Upon arrival in Hamburg, Amir had intended to enroll for the fall term in the graduate architecture program at the Hamburg University of Applied Sciences. To his chagrin, he was denied admission. The university said the program was full and could not accommodate him that term. Amir's father was outraged. He alleged the university was discriminating against his son because he was Arab. Amir threatened to sue for admission. The university quickly relented and he was accepted for the autumn term of 1992. Then, just weeks into his studies, he abruptly quit and enrolled in an urban planning program at a different school, TUHH, the Technical University of Hamburg-Harburg. He told his host family he had realized after starting classes the architecture program was insufficiently advanced; it would merely repeat what he had already learned in his undergraduate work in Cairo.

So he enrolled instead in TUHH's urban planning program. Technical University was south of the Elbe River in the old industrial suburb of Harburg. The Elbe forms what planners here call a cultural border. Technical University was built south of that border twenty years ago as an economic development measure for a declining manufacturing town. When Amir enrolled, the university was just ten years old and had only 5,000 students. It was not well enough known to have established itself as a destination for foreign students. In the autumn of 1992, there were fewer than one hundred students classified as foreigners in the entire university.[3] More than half of these weren't really foreigners at all, but ethnic Turks whose families had lived for years, often generations, in Germany but had never been granted citizenship. There were only about forty "real" foreigners among the students. Amir was one of a very few Arabs among them.

The planning program, from which he hoped to receive the German equivalent of a master's degree, was a good fit for Amir, in line with his analytical ability and meticulousness. The department was housed in a former police barracks, a plain, rectangular, wooden building left standing amid the swooping curves and startling metal sheaths of the new university that had been built around it. Fortuitously for Amir, the department's chairman, Dittmar Machule, was a Mideast specialist. Machule said he sensed in Amir someone who shared his passion for the ancient cities of the region. He described Amir as "tender, sensitive . . . he had deep, dark eyes. His eyes would speak. You could see the intelligence, the knowledge, the alertness."[4]

Hans Harms, another professor, said Amir was "almost shy in the beginning but engaged. I could see that he was listening, that what I said as a teacher would influence him." He was *"beeindruckt und beeindruckbar,"* impressed and impressible, Harms said.[5] Harms and Martin Ebert, a student who took several classes with Amir, recalled that Amir seldom jumped into discussions. He would sit and listen, often not saying a word, then come back a week later with a considered comment on the subject. Ebert said Amir wasn't much different outside class. He was careful about what he said, weighing it, never one to get excited. "I don't think it was possible to have a fight with him," Ebert said.[6] Harmut Kaiser, another classmate, said it was hard to draw Amir into political discussions in class, even when politics were directly relevant to the subject at hand. "He wasn't a guy who acted like he wanted to change the world—unlike a lot of other students in the group," Kaiser said.[7]

When students complained, as they do the world over, about a teacher's idiosyncrasies, Amir would join in the critique only if he thought a professor hadn't prepared properly or didn't know the subject. For those teachers who did, Ebert said, Amir showed a respect bordering on awe.

Amir showed somewhat less respect for his roommates. The difficulties he experienced when he lived in the small cottage with his host family repeated themselves in Harburg, where he moved into a university-subsidized apartment building called Centrumshaus. Each apartment had two bedrooms, a bath, and a kitchen. Amir lived in a third-floor flat at Centrumshaus from 1993 to 1998. He shared the flat successively with two men. In the end, Amir so aggravated both that neither could bear to be in his company. He seldom washed the dishes, they said, even if he had borrowed theirs to eat from. He almost never cleaned the bathroom. If asked, he would do it once, then not again for months. He left food uncovered in the refrigerator for weeks, affecting the taste of everything else. The shared kitchen was compact, functional, with a maple table that overlooked the street. It was a bright, sociable space, a place to sit for coffee or tea in the morning. Amir was often so inwardly focused he would walk in and out of the room without acknowledging anyone else in it.

His roommates grew to dislike Amir himself even more than the things he did.[8] The two men were very dissimilar. The first roommate was high-strung, anxious, the son of recent Asian immigrants. The second, a German, was laid-back and, after the failure of the first pairing, was chosen by the house manager specifically in the hope that he could get along with Amir.[9] It didn't work. Both roommates objected to the same personality trait in Amir: his complete, almost aggressive insularity. Just 5-foot-7 and wiry, Amir nonetheless had a heavy, foreboding presence. He was slightly awkward, stiff, and self-contained. His face, with its steeply angled planes and low, dark brow, lent itself to a variety of expressions that ranged from hangdog to menacing.

The first roommate tried to loosen up Amir, to include him in social settings. Early on, he invited Amir to accompany him and friends to a showing of Disney's animated film *The Jungle Book*. In the theater, Amir became so upset at the crowd's unruliness before the film began— described by the roommate as utterly normal conversation—that he seethed in his seat, muttering over and over again in disgust, "Chaos, chaos." He didn't speak another word during the film or on the walk home after. When they arrived back at the apartment, he stomped into his bedroom and slammed the door behind him.

Another time, Amir asked the roommate if he had any light reading material. Pleasantly surprised by the request, the roommate gave Amir a book of absurdist, Monty Python–esque short stories then popular among German undergraduates. He hoped it would entertain Amir and maybe become a shared interest. Amir took the book. The next morning, when the roommate came into the kitchen for tea, the book had been placed at the spot he normally sat at the breakfast table. There was never a word of explanation or thanks–not a word of any kind. "He was reluctant to any pleasure," the roommate said.

Amir had two credit cards, but almost never used them.[10] He spent very little money on food and very little time eating. When he did eat, he complained about the necessity of doing so. He generally ate alone. "We never shared food. We shared dishes. Mostly, he messed them up and I cleaned them," the first roommate said. "I remember sitting down at the table and Mohamed sighing, 'This is boring. Eating is boring.' He said it wasn't just that he wanted different food, it was just the act of eating."

Amir sometimes prepared a meal by boiling potatoes whole, scraping away the skins, then smashing the potatoes into a mound. He would eat his little potato mountain, without reheating it, for a week or more, sticking his fork into it and shoving the whole assembly back into the refrigerator when he finished what passed for a meal.

Each bedroom in the apartment was furnished with a bed, a desk, and shelves. The only thing Amir added was a slide-projector table. He had a camera and took photographs for his studies. It was the only attempt at a genuine creative activity anyone ever knew him to make.[11] More often, though, he used the table as a bookstand for his Qur'an. Amir prayed five times a day, fasted on holidays, and went to mosques whenever he could.

When he couldn't make it to a mosque, he prayed in his room, at work, even in the corner of classrooms. He almost always wore the same clothes: cotton slacks and wool sweaters, in particular a brown sweater-vest his mother had made for him, and a brown leather jacket in winter. His room was relatively neat, in large part because he owned so few things. He seldom wore shoes in the apartment, changing as soon as he came home into a pair of blue flip-flops. The second roommate said that by the end of three years, he and Amir were barely on speaking terms. Amir was so intense that the roommate once joked to his friends that he hoped Amir wasn't back at the apartment making a bomb to blow himself up with.

"In the end, I counted the days until Mohamed would leave the flat for good," he said. Students were allotted up to four years at Centrumshaus but could extend that to five if they were near graduation. Amir received the extension, much to the roommate's dismay. The roommate's girlfriend, a frequent visitor, was even more put off by Amir. He answered questions from her in curt, clipped tones and would never look her in the eye.

"It was a good day when Mohamed wasn't home," she said.[12]

The woman was so offended by Amir's attitude toward women that she conspired to get even with him. She persuaded her boyfriend to tack a postcard of a Degas nude in the bathroom above the toilet. The bathroom was small; a person couldn't open the door and avoid seeing the painting. Amir initially didn't respond to the provocation. Finally, three months later, he asked that it be removed. Then the girlfriend hung a poster in the kitchen, this one of the Muppet character Miss Piggy, dressed voluptuously in a negligee. Amir never said a word.

Dittmar Machule, the planning department chair, took a special interest in Amir. Machule was a committed Orientalist who saw his role at the university as both teacher and promoter of intercultural communication. When Amir early on chose the subject that would become the topic of his degree thesis—preservation of ancient cities in the Middle East—Machule was pleased.

"The other Muslim students, when they come to our world, they had problems with another cultural context," Machule said. "Either they try to get more and more a part of the Western culture, or they try to take something of that and this. . . . With Mohamed, I was somewhat impressed, I must say, with someone who didn't change, who tried to be as he was before, to try to learn, but to be who he was. I thought if this young man went back to his mother country, he could be able to work with the fundamentalist person, he could work with strong religious people because they believe in him."[13]

For several years, Machule had been conducting a research project in northern Syria, excavating the ruins of an ancient city near Aleppo. In 1994, he invited Amir to visit the site and consider Aleppo as the place to do fieldwork for his dissertation. Amir had already planned to join a summer excursion with other students to Istanbul, Turkey.

"I told him, 'Mohamed, try to come over to Syria; it's a direct bus line to Aleppo.' He arrived in August, early morning, after three days on a bus." Machule said Amir looked so rumpled and forlorn he couldn't help feeling sorry for him.

Amir spent several days with Machule at the excavation site, then said good-bye and went on to inspect Aleppo, which is one of the oldest continuously inhabited places on Earth. Throughout the developing world, in places like Aleppo, the collision of old and new wasn't merely theoretical. You saw it every day just by opening your eyes to the mudbrick neighborhoods, dense and jumbled and unchanged since before the Prophet's time. You could follow roads that Tamerlane had ridden en route to sacking the city in the fourteenth century, twisting along lines of elevation and drainage, losing yourself utterly in the ancient world, then suddenly come upon stark concrete apartment buildings that looked as if they had arrived intact from Moscow or Mars or a three-story minimall fresh off the boat from San Jose.

Amir visited a neighborhood called Almadiyeh Square. It, too, had suffered modern improvements. In the 1970s, the government had plowed broad new roads, improving access to and through the old town. Crews had cut part of a road right through the heart of Almadiyeh, tearing down what they needed, heedless of its value. They celebrated the completion of the project by erecting a small building next to it to sell souvenirs to the tourists the road was intended to carry.

"That was the only thing I ever saw him (Amir) get emotional about. He was very angry at the destruction of our old heritage," said Razan Abdel-Wahab, a Syrian engineer who worked at the Almadiyeh project.[14]

When Amir returned to Harburg, he told Machule that he would in fact make Aleppo the focus of his thesis. He and another student, Volker Hauth, made a second research trip to Syria at the end of the year. Amir was enlivened by the work, Hauth said. On a side trip to Damascus, Syria's capital, Hauth went to a mosque with Amir. Hauth was a devout Protestant and the two of them talked about religion often, but Hauth had never seen Amir in religious circumstances. At the mosque, he was surprised to see Amir leading prayers. Amir was self-assured, self-confident, and diplomatic. It was a revelation for Hauth, who knew the dour, introverted Amir from Harburg. Here he was a different person—looser, more talkative, animated, at times almost playful. It was as if he had been released, like "a fish in water," Hauth said.[15] Amir even made tentative advances to a woman he met in Aleppo. She teased him in return, calling him the Pharaoh.[16] He would later tell his father about his attraction to the woman, but said he couldn't pursue her because she was too forward, and thus not properly Islamic.[17]

Amir seemed to have many things working in his favor. After being forced by his father to leave home and go to Germany, an alien culture, he had found a structured system he enjoyed and work that engaged and challenged him. He flourished. He had gained a measure of acceptance and encouragement from his professors that he had never found at Cairo University. As an undergraduate in Cairo, Amir had never talked about his career, his dreams. Now he spoke of having found a future, about his desire to eventually return to Egypt "as an Arab to Arabia,"[18] to help build neighborhoods where people could live better lives.

OLD CAIRO

In the summer of 1995, Amir and Hauth won a grant from the Carl Duisberg Society, a German think tank that promotes redevelopment and preservation. Their grant allowed Amir, Hauth, and a third student, Ralph Bodenstein, to go to Cairo to study and analyze development plans the Egyptian government had devised for an old section of Cairo known as the Islamic City, a rich concentration of ancient monuments, modern marketplace, and medieval architecture. Both the glory of Egypt's history and its passing are nowhere more evident than in the Islamic City, which is guarded by a high, imposing stone rampart that stretches between the Northern Gate and the Gate of Victory and is packed with mosques and churches and even a synagogue. History is layered along the crooked streets and half-lit alleys lined with bins of lentils, dry noodles, fresh peppers, acorn-sized lemons, and dark olives the size of Ping-Pong balls.

What the three young architects found appalled them. The government planned to "restore" the area by removing many of the people who lived there, evicting the onion and garlic sellers, repairing the old buildings and bringing in troupes of actors to play the real people they would displace. The old quarter sits just west of the sprawling City of the Dead, Cairo's central cemetery, which is home to centuries of the city's deceased as well as thousands of squatters who live in and among its burial crypts and mausoleums. The project engineers were already at work evicting the residents of the cemetery, dead and alive. The earthmovers cut swaths along the edge of the city of the dead, scraping away another meter of earth and the crypts and bones underneath it. They were "fixing" the thousand-year-old cemetery.

The three young architects discussed the huge flaws in the project, then tried to express their concerns to the people running it. Bodenstein described what happened: "We had a very critical discussion with the municipality. They didn't understand our concerns. They wanted to do their work, dress people in costumes. They thought it was a good idea and couldn't imagine why we would object."[19] It was Amir's first sustained professional contact with the Egyptian bureaucracy, and it distressed him, Bodenstein said. "Mohamed was very, very critical of the planning administration, the nepotism. He had begun to make inquiries about getting a job after school, and he had difficulty finding anything. He did not belong to the network, where jobs were handed down from one generation to the next, to political allies. Mohamed was very idealistic, humanistic; he had social ideals to fulfill."

Amir's complaints about the difficulty of finding a decent job were not unique. Egypt's ambitious, virtually free system of higher education pumped out many more graduates than the economy could handle. His alma mater, Cairo University, produced more than 1,000 engineering graduates every single year. The net result was the more education you had, the less likely you were to find a suitable job. In one recent year, young Egyptians with graduate degrees were 32 times more likely to be unemployed than illiterate peasants. [20]

Bodenstein said Amir's critique of the government grew more expansive as the study project went on. Amir worried that the redevelopment would turn the old city into an Islamic Disneyland. Such Western influences, he said, were the result of Mubarak's eagerness to curry favor with the United States. Bodenstein felt there was more at stake for Amir than the redevelopment project: "There are many things that might have contributed over the years, contributed to a certain embitterment he was feeling. I don't think it was religious. Religion provided the vocabulary, not the cause. The cause was political."

The study project lasted five weeks. When it was done, Hauth and Bodenstein returned to Hamburg. Amir stayed on in Cairo and spent time with his family, which had moved from Abdin west across the river to Giza. Amir went back to his old neighborhood to visit. He asked his neighbor, the mechanic Khamees, to check out his worn-down Fiat. Khamees said they talked about old times while he inspected the car. As they talked, the afternoon call to prayer sounded. Amir excused himself to answer it. It was the first time Khamees had ever seen anyone in Amir's family go to mosque.

Religion had become a chief focus of Amir's life. With his father's blessing, Egypt Air contacts, and financial assistance, he joined that year's pilgrimage to Mecca, in Saudi Arabia, an important, often powerful experience in a Muslim's life. Every believer who is able is supposed to make the journey at least once. Out of necessity, Saudi Arabia restricts the number of pilgrimage visas it issues, otherwise the holy sites would be overrun. As it is, more than a million people make the pilgrimage, or *hajj*, as it is called, every year. The visas are highly prized and to be allowed to make the pilgrimage at such a young age—Amir was 27—was a privilege.

When Amir returned that winter to Harburg, Hauth thought he was, if possible, even more quiet, more introverted, and more fervent in his religious practice.[21] John Sadiq, a classmate who also worked with Amir at a part-time job, saw the same change.[22] Amir told Hauth that he eventually wanted to return to Egypt to work as a planner but despaired of a political situation in which religious fundamentalism was regarded as a threat to the government. "He lived in fear of being criminalized for his religious beliefs," Hauth said.

Amir never had many German acquaintances, and none who regarded themselves as close friends. One reason was his introversion. Another was his narrow range of interests. Amir was an exceptionally resolute, disciplined, stoic man. He owned almost no books, didn't like food, and didn't listen to music. As far as anyone knew, the only movie he had ever seen was the one his first Harburg roommate had dragged him to, which, of course, he hated. He simply wasn't much fun to be around unless you wanted to talk about Islam, Cairo, or city planning, not subjects known to foster friendships among a cohort of Germans termed Generation Golf (after the Volkswagen car, not the sport) and widely regarded as the most materialistic in German history. The German student culture in which Amir lived was drenched in beer and concerned more with football rivalries than politics or religion.

During his early years in Hamburg, Amir worked part-time as a draftsman at an urban planning firm, Plankontor. He was an excellent employee, said Jörg Lewin, one of the firm's partners.[23] Helga Rake, another partner, called Amir *kleinteilig*, meaning he paid precise attention to the smallest details without seeing the broader picture.[24] "I think he embodied the idea of drawing," Lewin said. "'I am the drawer. I draw.'"

Amir went out of his way to visit project sites to get a better feel for what he was drawing, something far beyond what was required or what others did. Plankontor was a small company located in one of the more fashionable neighborhoods of Hamburg, near trendy restaurants and an art-house cinema. It was a casual place, hip in its own way, and, like many such firms, prided itself on egalitarian relationships among employees and between the workers and their bosses. It was relaxed and companionable. Far from disdaining Amir because he was Arab, the firm's employees seemed to take pride in their inclusiveness and were puzzled by Amir's rejection of it. Though well-regarded and even liked, he never joined the Plankontor family. The firm always invited him on its regular holiday trips; he never went. Coworkers asked him to lunch; he declined. He wasn't exactly impolite. He just said no to all offers of socializing with coworkers. He stayed at his drawing table and worked, or, when it was time, knelt beside it and prayed. He was—particularly for a university graduate student—enormously respectful of authority. He did what he was told and did it, Lewin said, with extraordinary single-mindedness. Although already a trained architect and a prospective city planner, Amir—in four years at the company—never once offered an opinion of the plans he was asked to illustrate. He was assigned to make maps; he made maps.

CHAPTER 3

Friends

BY 1996, AMIR HAD MET MOST requirements for his graduate

HAMBURG

degree. Mainly what he had left was to write a dissertation.¹ But after he
returned from making *hajj*, he seemed to lose interest in his academic
work. He grew more and more immersed in religion and became a well-
known figure at local mosques. He grew a beard—a sign of devotion to a
fundamentalist interpretation of Islam. He eventually developed a circle
of acquaintances from Al Quds, the most radical mosque in Hamburg,
and other mosques in the city. He saw less and less of his German colleagues. Volker Hauth, who had been a study partner for more than two
years, didn't see him at all for the next two years. Amir's new friends
called on him frequently at Centrumshaus. He sometimes invited groups
of them to dinner. He made soup.² Other than answering the doorbell,
there was no interaction between Amir's friends and roommates. He never
even made introductions.

Notwithstanding that they were all men, and all Muslims, the friends
were a varied group. They included North Africans, Gulf Arabs, German
Muslims, Syrians, even Indonesians. These new friends were mostly students, but also older men, including some who had been in Germany for
decades and who functioned as informal guides to German society to
many of the younger, more recent arrivals. They helped the newcomers
find jobs, schools, and, in some cases, causes to believe in. Amir was

unusual among the more recent immigrants in that he spoke perfect Arabic and passable German and English. Many of the younger men spoke none of these languages particularly well.[3] Amir became a go-between. He helped new arrivals find apartments and in general helped them negotiate the German bureaucracy, which was quite generous if you minded its many rules. Amir was older than most of the students, so he served as a bridge between the generations as well.

Many of the older men in the circle were part of a generation who had come to Germany in the 1970s and 1980s. In particular, there were a number who had left Syria after President Hafez al-Assad (father of the current president, Bashir al-Assad) declared war against Islamic fundamentalism in 1982. Many of the Syrians had been members of the Muslim Brotherhood, the protorevolutionary group founded in Egypt in the 1920s, and out of which much of contemporary radical Islam springs. Political Islam had not suddenly been invented by the Brotherhood. From the prophet Mohammed forward, politics and Islam had been intertwined beyond any ability to unravel them, but political activism among the faithful had been quiescent for centuries. Civil discourse within the Arab world was largely a one-way conversation—from the top down. The Brotherhood challenged the state's monopoly on political discussion, and many of them paid for that challenge with their lives. Others fled.

For thirty years there had been a small but persistent number of politically active Muslims in Germany. One study identified 58,800 members of "foreign extremist organizations."[4] Islamic extremists predominated; most were Turkish or Kurdish. About 2,300 were Arabs, most of those associated with various Palestinian groups and issues.[5] The Brotherhood exiles who fled the Middle East in the 1980s were Europe's first generation of pan-Arab Islamists devoted to a broader cause. They had been energized by the holy war against the Soviet Union in Afghanistan and maintained a loose network across the continent. It would later seem to some investigators that these activist networks were part of a great conspiracy because they all seemed to know one another. In fact, many of them did know one another, but this could have been as much an artifact of community as conspiracy. It often was impossible to tell the difference.

Just as they did among apolitical immigrants, and for the same reasons of colonial and cultural history, different ethnic groups tended to cluster in different countries in Europe: The French Islamists were overwhelmingly Algerian; Tunisians made up a large portion of those in Italy; London had long been home to a significant number of Gulf Arab and Egyptian expatriates and Britain served as the spiritual and intellectual capital of European fundamentalism; Spain, like Germany, was home to a cadre of Syrians, but the overwhelming majority of its activists were Moroccan; the Netherlands and Belgium had both Algerians and Moroccans. Members of these ethnic groups often knew their peers in other countries. They shared educational and recruitment materials. Speakers and preachers roamed freely across the continent. Some of the groups built sophisticated fund-raising apparatuses and those along the Mediterranean operated rest and recuperation centers for men who fought for the Muslim causes in Afghanistan, Chechnya, and the Balkans.

Some of these groups were under prolonged and intensive police surveillance. The Spanish, French, and Italians, in particular, kept close check on those they considered extremists. Some Italian and Spanish fundamentalists were under police surveillance for half a dozen years. In most places, Islamist activities centered around mosques, as they did in Hamburg.

Amir began teaching informal religion classes to younger Muslims at Al Quds and at a Turkish mosque near the university in Harburg. The classes mixed religious instruction with political discussion. The plight of the Palestinians was a frequent topic. As was typical of him, Amir was quite stern with his students.[6] He criticized them for wearing their hair in ponytails and gold chains around their necks and for listening to Walkman music players. Music was a product of the devil, Amir said.[7] When young women showed up for one of the classes, Amir sent someone to tell their fathers they were unwelcome and should not return.[8] Most of the students decided he wasn't the sort of teacher they wanted and dropped out fast. Over time, as many as eighty students started his classes. By the time Amir left Hamburg, he had only a handful left.[9]

Even when he wasn't in the classes, Amir thought about Islam all the time, said Ahmed Maklat, a TUHH student who joined one of Amir's prayer groups.[10] Amir frequently stopped by the student apartments where

Maklat and Mounir el-Motassadeq, a Moroccan engineering student Amir had met in the mosques, lived in Harburg and walked with them to and from prayers. He talked about the meaning of the prayers in both directions and instructed Maklat in how they should be said.

"Don't cut corners when you pray," Amir said. "Pray the last three of the four prayers separately, not en bloc. Also, usually you pray the duty prayer, a special prayer, and you also praise Allah with a third prayer, which is not a duty." And on and on and on. Maklat started leaving the mosque early, before prayers ended, so he wouldn't be subjected to Amir's critique on the way home. He began referring to Amir as the ayatollah.

The instructional work was time-consuming. Amir taught his own classes on Friday and Sunday evenings, took part in a discussion group led by another man on Saturdays, and was coleader of yet another group that met Tuesdays and Thursdays.[11] That group was led by a German convert who disliked Amir's rigidity and began fining him whenever Amir, who had to come by train all the way across town, showed up late.[12]

Another man who taught occasional classes became one of Amir's best friends. His name was Ramzi bin al-Shibh, although most people knew him simply as Omar. His real name, he said, had no religious meaning, so he adopted the name of the prophet Mohammed's successor, the second caliph of Islam. Many acquaintances in Hamburg didn't even know Omar had another name.

There was a second reason Omar used that name. It was the identity he had adopted when he first arrived in Germany, claiming to immigration authorities to be a university student from Sudan fleeing persecution. He made up an elaborate story about stowing away on a Sudanese ship, sleeping on the floor of the crew cabin during the sea passage. The next month he appeared before immigration authorities in Hamburg and handed them a note:

"Dear Sirs, with this application I ask you for political asylum. Yours sincerely, Ramzi Omar."[13]

Hamburg authorities, who had more asylum candidates than they could handle, gave him a train ticket to Lübeck, north of Hamburg, and instructed him to make his application there. Omar told authorities in Lübeck that as a first-year economics student in Khartoum he had taken part in an antigovernment demonstration and been arrested. The government later broke into his apartment and stole his identity papers, he said. "There is no freedom in Sudan, no human rights, no respect for human beings," he said.[14]

The Germans accepted his application, put him on a waiting list for a court hearing, and gave him temporary quarters in a series of refugee camps maintained specifically for asylum seekers.

Omar was playing a scam and was hardly alone in it. Throughout the 1990s, Germany received on average more than 100,000 asylum seekers a year, fewer than a tenth of whom were able to demonstrate sufficient proof to the German government they were fleeing anything. What asylum seekers wanted most was access to Germany's generous welfare apparatus, which would provide them with free health care and money for food and lodging, virtually forever. Gaining asylum would also give them political status and access to a Europe-wide visa. Asylum, if approved, was a virtual ticket to a new life.

Omar had actually arrived in Germany a month earlier than he told authorities, in August 1995, having transited through the United Arab Emirates and Cairo.[15] He wasn't Sudanese at all, but Yemeni and had never been politically active or attended university anywhere. He couldn't afford it. He worked as a messenger boy at a bank in Sana'a, the capital of Yemen, where he lived with his mother and two older brothers. His father had died nearly a decade before. The father's death and the relative poverty the family endured after had greatly complicated Omar's lifelong ambition—to go to the United States and study computers.[16] He came instead to Germany.

SANA'A

Yemen occupies the southwestern corner of the Arabian Peninsula. It's where the great 1,000-mile run of desert finally ends at the base of a series of ancient volcanic ranges that divide the sands to the north and the endless seas to the south. In the valleys and plateaus between the ranges is one of the oldest civilizations on Earth. The country that arose from that civilization is beautiful, backward, terribly poor, proud, and biblically ancient. The Queen of Sheba ruled most of it ten centuries before the birth of Christ; Noah is said to have built his ark at a harbor on the southern seacoast. Whether or not it was swamped by the raging waters Noah sought to avoid, the country has suffered more than its share of calamities since. The latest, and in some ways greatest, of these was a prolonged civil war, beginning in the 1960s and continuing in some form or another for thirty years. It was in part yet another proxy fight between the Soviet and American empires and left the place flooded with weapons, hatreds, and instability. Yemen has 18 million people, 60 million guns, and a central government still struggling, and failing, to assert control. "The fight was gone, the weapons remained," as one man put it.[17]

The main divide in the country is between the north, where Sana'a, the capital is located, and the south, where the port city of Aden was the capital of a separate republic—the People's Democratic Republic of Yemen—for most of the last fifty years. The north and south were finally unified, at least nominally, in 1990, but much of the country remains a collection of tribal fiefdoms and tiny outlaw empires that resist all authority but their own. This is in keeping with the history of the place: although there is a clear sense of national identity, it has never been unified. That disunity has become, in many ways, Yemen's defining characteristic.

Ramzi bin al-Shibh was born in the central Yemeni province of Hadramaut, in 1972. Hadramaut is legendary for the mud-brick skyscraper cities in its remote interior valleys, which were once the center of the world trade in frankincense. The incense, which was used in religious ceremonies and for embalming for centuries, made Yemen rich, but the trade died in the nineteenth century and not much good has happened there since.

Ramzi's family lived south of the mountains in the farmlands of the southern coastal plain. The father, Mohammed, was a merchant who moved the family to Sana'a when Ramzi was just a boy. Sana'a, at least in comparison to the rest of the country, is a modern city with international air connections, hotels, and a bustling if dismal city center. The streets are full of ancient Peugeots, Toyotas, and Nissans, patched with duct tape. People who can't afford cars, which is most of them, ride in share-taxis, open-ended minivans overflowing with as many as a dozen passengers.

The city sits on a volcanic plateau 7,000 feet above sea level, with less heat and more rain than anywhere else on the Arabian Peninsula. Men wear long plaid skirts and western style dress suit jackets, often with a manufacturer's label sewn on the exterior of the left sleeve just above the wrist. They sometimes add a red-and-white checked scarf, worn as a shawl. Most wear daggers with long curved handles—10 to 12 inches is standard. Men and boys walk the streets with their hands resting on their dagger handles, as if ready to draw the knives at a moment's notice. The sheaths for the knives are worn at the middle, rather than the side, of the belt, and the dagger handles point up and out from the belt, like saddle horns. The women, on the rare occasions they're seen in public at all, wear black, head to toe in the Saudi style, although you sometimes catch a glimpse of blue denim below the bottom hem.

Alone among the countries of the Arabian Peninsula, Yemen has no discernible oil reserves. The economy has been stagnant for years and there is little real work. In the afternoons and through the evenings, small groups of men sit on doorsteps and street corners, or just crouch on the side of the road, chewing *qat*, a narcotic plant that is grown in the interior. The men start chewing around lunchtime and some spend the rest of the day at it, squatting around shared tin plates of pureed chickpeas.

The *qat* economy has reversed the normal urban-rural distribution of wealth in the country, enriching villagers who grow the plant and impoverishing the presumed cosmopolites of the cities who consume it. The drug is a mild stimulant, the chief effect of which is sleeplessness. Men who chew it through the afternoons frequently can't sleep until far into the night, causing many to nap through the working hours of the morning.

What work there is tends toward small-scale retailing. Armies of peddlers occupy street corners, hawking everything from facial tissue to cut-glass chandeliers. Boys push wheelbarrows full of turmeric and whole tobacco leaves. Small shops sell bananas, mangoes, used watches and cell phones, spices, and, of course, daggers. Men spend what seems to be hours perusing racks and racks of knife handles, studying differences no one but a Yemeni could see, much less appreciate. Women spend equivalent hours looking at the same endless variety of all-black clothing.

The Shibh (pronounced *shay'bah*) family lived in a small house in Hasaba, just east of the central city. The neighborhood is working-class, filled with gray mud houses built in the distinctive vertical style of the country, two, three, and four stories tall. Holes for windows are punched irregularly across the façades, making the buildings look from a distance as if they have suffered an artillery barrage. There's a neighborhood mosque, a round mud watchtower, and an endless skein of crooked alleys.

When the father died of illness in 1987, Ramzi's oldest brother, Ahmed, took charge of his upbringing. Ramzi was a mischievous child, always in one sort of small trouble or another, but he got away with it because he was so happy and open. "Mother loved him so much. He was the favorite," Ahmed said. "We used to ask our mother, 'Why do you love Ramzi more than us?' She couldn't say, she'd just smile. He brought gifts for her—winter clothing, jewelry, medicine for her teeth. He was a very good boy."[18]

As an adolescent, Ramzi was the favorite uncle of his younger cousins. "He was always the one who would play with the children. They loved him," Ahmed said.

The family was not particularly religious, and except for Ramzi, seldom went to prayers at the mosque. Ramzi, though, took his Islamic studies seriously and for a time, helped at the local mosque school, teaching the Qur'an to young children.[19] He even moved into the mosque for a period of months.

This was at a time when Yemen was being roiled by radical Islam. During the great Islamic insurrection in Afghanistan, Yemen, even more than most Arab countries, heartily endorsed the Muslim mujahideen rebellion against the Soviet Union, and thousands of Yemeni men and boys went off to help fight it. Yemenis have had a reputation as fierce fighters from the time of Mohammed, and the culture included an idealized notion of Yemeni men as the true Islamic warriors.[20]

"Imams preached it. The government encouraged it," said Hussam Sanabani, a Yemeni who came of age during the jihad against the Soviets. Even high school students would go off during summer vacations. "All of a sudden they wouldn't be here anymore," Sanabani said. Many of these men had their way paid to Afghanistan by Saudi charities and religious leaders. This happened throughout the Arab world, but nowhere to the extent it happened in Yemen, which was the poorest of Arab countries, the least Westernized, and the one with the longest, deepest warrior tradition. What distinguished Yemen even more was what happened afterward, when the jihad was over. Many Arab countries were glad to send men off to fight in the war. They were only unhappy when they came back. The jihad burned too hot at home. Men fired by its call to action often could not cool down when they returned. Many countries did all they could to keep them from coming back. Yemen was the opposite. It welcomed home its own Islamic warriors and everyone else's as well. Yemen opened its borders to every jihadi who had nowhere else to go.[21] Visas were issued without prior notification upon arrival at the Sana'a airport.[22] All you had to do was show up and claim one. The country became home to thousands of homeless revolutionary Islamic warriors, many of whom had nothing to do but plot further revolution.

Religious schools were founded with money provided from the oil-rich Gulf states, who were eager to spread their version of Islam. One school, in particular, Iman University, became famous for the firebrand radicalism taught by its founder, Sheikh Abdul Majeed Zindani, who had fought alongside and befriended Osama Bin Laden during the Afghan war. The school was funded largely by Gulf State charities. His rhetoric was by any outside analysis remarkably hostile to the west, but it was unexceptional in the widening circle of Islamists.

Ahmed bin al-Shibh was busy trying to raise his father's family and soon enough his own, too. He went to work as an analyst at an economic research institute. This new radical Islam passed him by entirely, he said. Ramzi, though devout, seemed unaffected, as well. He took a part-time job as a clerk in the administrative section of a bank. The job title was *kattib*, which means writer in Arabic, but was used colloquially to describe someone who carried other people's writings—that is, a messenger boy. He started the job during one summer vacation in high school. "He liked it and stayed after. He wanted to help the family. And he was trying to collect money to support himself," Ahmed said. Ramzi continued in the job past graduation, but as he got older he became a less than reliable employee, showing up late and arguing with superiors about his work.[23]

He enrolled briefly in a business school but planned for several years to go abroad. "He was always one of those guys talking about studying in Europe or the U.S.," Ahmed said. "He saved money for it for years. He wanted to improve himself. He was very ambitious and forward-looking. We couldn't afford the expense of sending him.

"He had to do it by himself. . . . He didn't always have in mind to go to Germany to study. Britain or the United States were first. He had his colleagues at the bank who went to the United States to study computers. He felt he should study like them, but no visa."

In the end, Ramzi told his family he had decided to go to Germany to study economics and politics. He never said why he chose Germany. Ahmed said he thought it was the one place for which he was able to get a visa. Ramzi arranged all the details himself.

THE CONTAINER

Asylum seeking was widespread throughout western Europe in the 1990s. In 1996, German courts heard 194,451 asylum cases. There was a lot to run from in the world and Europe had been a particularly hospitable place to run to. But it had gotten to be burdensome for the destination countries and politicians were being pressured to rein in their hospitality. German politicians were not excepted. They had enacted a law that provided for immediate expulsion of asylum seekers if they arrived by land, the underlying assumption being that anyone already in any country that bordered Germany wasn't fleeing *from* persecution, but *to* Germany's generous welfare state. The change in the law was no secret among the asylum seekers themselves, and after it was enacted they began telling authorities they arrived by airplane—in which case they were asked to provide carrier information—or ship, which provided an alibi that was conveniently difficult to verify. Shibh's claim of stowaway passage on a Sudanese ship both averted automatic expulsion and was impossible to disprove in and of itself.

His initial claim was nonetheless denied mainly on the grounds of implausibility. He appealed and was assigned to a camp just north of Hamburg in a town called Kummerfeld, or Field of Sorrow, to await final disposition. Kummerfeld was hardly a camp at all. It consisted of a single building about the size and shape of a ship container and, in fact, that's what everybody called it—the container camp. The building was divided into three sleeping rooms, one bathroom, and one kitchen. It was cramped, drafty, and entirely unpleasant.

Container residents were paid modest stipends. They were encouraged to find work and, if they didn't, to work for the city doing various manual chores. The system was not rigorous, however. Typical of Germany's modern bureaucracy, so long as the asylum seekers showed up for weekly roll calls, they were free to come and go as they pleased and also free to pick up approximately $500 in monthly welfare coupons. Still, few people liked living in the camps. Residents complained constantly and hardly anyone stayed there during the day. Some of the camps, including Kummerfeld, became notorious among law enforcement authorities as havens for drug sales and distribution points for false documents.[24]

Shibh had come to Germany prepared. He was registered as a resident at the apartment of a fellow Yemeni, Mohammed bin Naser Belfas, six months before he ever arrived in the country.[25] How he established contact with the man was unclear, but Belfas guided Ramzi's initial preparations for coming to Germany, according to Ramzi's brother, Ahmed. Belfas was an unusual fellow in his own right. He was born in Indonesia to Yemeni parents and spent part of his childhood in Yemen. He went to university in Cairo and came to Germany on a six-month tourist visa in 1972, then stayed thirteen years before he was discovered to have overstayed the visa and briefly jailed. When he was released, the Germans intended to deport him—but to where? By that point he had lived in Germany longer than he had lived anywhere else; he had no legitimate passport and there was no place to deport him to. The government finally relented and allowed him to stay in Germany. He was given work permits and eventually found a job on the night shift at a suburban Hamburg postal facility. He was granted citizenship in 2000.

Belfas was unmarried and, by the time Shibh arrived, middle-aged. He's a squat, open-faced man with a fringe of whitening hair around his balding head and a short gray-white beard. He devoted almost all of his nonworking time to Islamic causes and was well known among Muslims throughout Germany. Friends likened him to a lay missionary who made it his task—one called it a mission—to unite the varied ethnicities and sects of Muslims in Germany.[26] Belfas's apartment, which was usually a mess, functioned as a sort of Islamic lending library. People came from all over the city to borrow his books.[27] He traveled the country, taking the train or, more often, cadging rides from friends. He spoke wherever anyone would listen, in colleges or mosques, sometimes to groups no bigger than a handful. All of this travel and night work must have worn him out. He almost always came late to Friday prayers at Al Quds and was notorious for sitting next to the wall, then dozing off against it.[28]

Shibh was registered as a resident at the container camp for more than two years. He was among a group of Arabs camp administrators called "the bulls" because they were aggressive in complaints about the living conditions. Shibh was the tamest of the bulls, "always in the background. When there was trouble he would disappear," said Michael Hirsekorn, the camp's administrator.

Shibh seemed intent on escaping the container. He worked numerous jobs, sometimes two at a time. He baked pretzels and gardened at an old folks' home; he worked as a janitor and got in trouble for taking a job at a Chinese restaurant.[29]

Shibh's asylum appeal was eventually denied. The judge said he doubted Shibh was even Sudanese, much less fleeing Sudanese persecution.[30] The judge, of course, was right. The dismissal of the claim had little effect. Shibh had been anticipating losing the appeal and prepared other means of getting his visa. He had told his family he had come to Germany to study, but he never pursued it with any diligence. Instead, he registered several times at different kollegs, which offered entry-level language classes for foreign students. He was accepted for enrollment at the University of Applied Sciences in Wismar, northeast of Hamburg on the Baltic coast. But he never showed up for more than a handful of classes.[31] Had he simply wanted to learn German he could have done it at the container camp, where classes were offered regularly. He never attended. His real intent seemed to be to secure documentation that would allow him to be granted a student visa. Even before his asylum claim was finally decided, he had obtained the student identification papers and returned to Yemen. He also had with him a letter from a businessman in Hamburg, Yusef Masoud Ahmed al-Sayyid, stating that Sayyid would pay Shibh a stipend of $400 a month while he studied. Shibh never actually held the job, nor apparently intended to. He was just ensuring that he would be able to stay in Germany if he lost his asylum case. He took the letter and the college documents to the German embassy in Sana'a, where, using his real name rather than the Omar identity, he received a legitimate German student visa and reentered the country under a new identity, albeit his own—as Ramzi bin al-Shibh.

Shibh stayed off and on at Belfas's apartment and used it as a mailing address for years. He seldom had a fixed address of his own. All he needed, a friend said, was a mattress and a corner to put it in. In a way, Shibh eventually came to emulate Belfas, devoting himself entirely to the cause of Islam. He seldom held a job for more than a week or two; as a result, he always owed money to somebody.[32] He traveled constantly, meeting Muslim activists throughout Germany and elsewhere in Europe.

At the very beginning of this journey toward activism he met Amir. Within weeks of arriving in Hamburg, Shibh had found his way to Steindamm. He shopped at an Arab grocery there,[33] frequented a restaurant, and, presumably, attended prayers at Al Quds, which was just across the street from the restaurant. If he was in the company of Belfas he could have been introduced to almost anybody, since Belfas knew almost everybody. Belfas, for example, knew Amir through an Egyptian student who had studied German with Amir in Cairo. However it happened, by the winter of 1995, Amir and Shibh had met; they quickly became close enough that Shibh lived in Amir's student apartment while Amir was away on winter break.[34]

CHAPTER 4

Pilgrims

HARBURG

AMIR, AS HE PROGRESSED AT TUHH, continued living at his university apartment and working at Plankontor, the planning firm. Plankontor's business declined in 1996 and the firm cut back. Amir was laid off. He hated to lose the job, he told his boss, Jörg Lewin, but left gracefully.[1] With characteristic diligence, he even sent back money he thought the firm had overpaid him on his final paycheck. Amir finished his course work at TUHH the next spring. All he had left to get his degree was to write his thesis. Instead, he vanished from school. He had almost no contact with the university for nearly a year beginning in the fall of 1997.

Amir was gone from Hamburg for part of the period but mainly he disappeared into the city, not away from it. He taught a series of seminars for the think tank that had sponsored his research trip to Cairo. The seminars were for students undertaking similar projects. Amir wasn't markedly different as a teacher than he had been as a student. He was well-prepared, thorough, unexciting, and serious, said a man who attended one seminar.[2] The man, an Egyptian, said that he was initially excited to meet another Egyptian so far from home, but that Amir, while not hostile, showed little interest in personal conversation. The seminars included evening social events. Amir attended none of them. The Egyptian student said Amir always seemed preoccupied. There was "a wall" between him and the students, he said.

Shibh, or Omar, as everyone knew him, had no such problems. Where Amir was hard, he was soft. Where Amir was focused, dark, and stern (iron, as his former landlady described him), Omar was dreamy, light, and happy-go-lucky. "He was very strict when it came to his faith, but he stayed relaxed, anyway . . . he had a more relaxed approach than, for example, Amir, who always seemed to be strained," said Shahid Nickels, a young South African Muslim convert Omar befriended at Al Quds.[3] He was "in love with life" and saw himself as a fighter, Nickels said. Like many Yemenis, he raved about the physical beauty of his home country, especially about the mountains of Hadramaut.

"Omar was very religious, very charismatic, very charming, very intelligent. You had to like him," Nickels said. "He never criticized people directly, but would speak about similar mistakes he had made. His emotional intelligence and his knowledge of human nature were admirable; you could see that he always understood the motivations of people he met. He knew how to sell himself and to represent himself."

Omar's physical bearing reinforced his personality. He was a small, thin man with an open, smiling face, a deep brown complexion, and a soft, high voice that never seemed threatening. He held hard beliefs but presented them softly, Nickels said. He was rarely downcast, but when he was, he placed his faith in the hereafter. "What is this life good for? The paradise is much nicer," he said.[4]

Nickels was welcomed into an informal group drawn from the university and the mosques that began coalescing around Omar and Amir in 1997. The group consisted mainly of younger men, many of whom had less than comprehensive knowledge of Islam, or, as in the case of Nickels, almost none at all. Omar was the first person to tell Nickels the elementary fact that his first name, Shahid, meant "martyr" in Arabic. (Nickels's parents just liked the sound of it.) Omar and Amir functioned as mentors for the younger men, most of whom were students. Amir and Omar didn't proselytize so much as court other Arabs. They showed up at mosques and study groups around town and offered to help translate and explain passages of the Qur'an.[5] The group grew to include Mohammed Ragih—like Omar a Yemeni—and several of the young Moroccans from Al Quds Mosque—Zakariya Essabar, Mounir el-Motassadeq, Abdelghani Mzoudi, and Said Bahaji. Motassadeq and Mzoudi had known Amir at least since early 1996, when both had witnessed a will he drew up.[6] Essabar was a more recent arrival and Bahaji, whose mother was German and who had been born and spent his early childhood in Germany before his father took the family back to Morocco, had just recently returned.

It is hard to appreciate how much time these young men spent thinking, talking, arguing, and reading about Islam. It became for some of them nearly the only thing they did. Some of the men in the group later met frequently in the backroom of a bookstore near Al Quds. There, they listened to audiotaped sermons and jihad chants. One sample lyric: "When I die as a martyr, I die as a better human being."[7] They heatedly debated the meaning of Qur'anic texts or words of the Prophet. Almost all of their conversation was about religion. Said a German investigator: "They are not talking about daily life stuff, such as buying cars—they buy cars, but they don't talk about it, they talk about religion most of the time . . . these people are just living for their religion, meaning for them that they just live now for their life after death, the paradise. They want to live obeying their God, so they can enter paradise. Everything else doesn't matter."[8]

BEIRUT

The group was fluid. People drifted away, others—frightened by the intensity that was developing—fled. Some were so fearful they moved to other cities. Over a period of three years, a couple dozen men came in and out of the group. One of the recruits was a young Lebanese man named Ziad Jarrah, who was by any accounting an unlikely candidate for Islamic warrior.

Jarrah was the handsome middle child and only son of an industrious, middle-class family in Beirut in 1975. His father was a midlevel social service bureaucrat, and his mother, who came from a well-off family, taught school. They drove fashionable Mercedes automobiles and owned an apartment in a densely populated Sunni area of the city. Today, the neighborhood is a picture postcard of compacted urban living. Laundry hangs out to dry from balconies eight, nine, ten stories above the street. Young boys kick balls between lampposts draped with the flags of local football clubs, and men in sleeveless T-shirts hunch over small backgammon boards on the sidewalks. It wasn't such a friendly place when Jarrah was growing up. The Jarrah home was just two blocks from the Green Line that divided combatants during Lebanon's long civil war. Two blocks the other way was the sprawling Sabra Refugee Camp, where hundreds of Palestinians were slaughtered under the occupying Israeli Army in 1982. It has since become a tightly packed permanent Palestinian settlement.[9] Many buildings in the neighborhood are still pockmarked from small-arms fire. Others bore greater damage from artillery and mortar shells; here and there buildings are still missing entire facades.

At the height of the war in the late 1970s, the Jarrah family bought a second home as a refuge in Marj, the Jarrah clan's ancestral village in the Bekka Valley. Samir, the father, spent most of his time at the country house. Ziad was a city boy, though, and seldom went out to the valley, especially after relative calm returned to Beirut in the 1990s. Ziad attended private Christian schools—a mark of relative affluence and aspiration, not religious inclination. He was bright, but didn't apply himself much at school, at one point nearly flunking out of high school.[10] He was, the family said, more interested in girls than geometry.

The family was secular Muslim, which was the norm in Beirut, a city that prides itself on its cosmopolitan outlook. Ziad's uncle, Jamal Jarrah, said the family was so lax about religion that he, Jamal, was nearly a teenager before he even knew he was a Muslim.[11] His mother's best friend worked for a local Protestant minister, and she took Jamal to church with her every Sunday; he grew up thinking he was Christian. Ziad's father was not religious, either. The household was easygoing—the men drank whiskey and the women wore short skirts about town and bikinis at the beach.

Jarrah was bright, but inattentive in high school. When he graduated, his family gave him the choice of attending university abroad in the two places the Jarrahs had relatives—Toronto, Canada, or Greifswald, Germany, a tiny northeastern backwater on the Baltic coast. Jarrah was not prepared to go as far as Canada, so chose Germany. He and a cousin, Salim, arrived in Greifswald in the spring of 1996, not long after a vivacious young woman named Aysel Sengün had enrolled there in the College of Dental Medicine.

Ziad and Aysel met within a month of Jarrah's arrival, on the day he moved into his student quarters at the University of Greifswald. Aysel lived just down the hall. She was the daughter of conservative, working-class Turkish immigrants to southern Germany and had been in Greifswald for a semester. She already had a boyfriend, but Jarrah must have seemed an answer to many dreams: a big-city boy with an easy smile, like her a moderate Muslim who enjoyed a good time. She wondered about potential problems, confiding to her sister that Arab men could be domineering, but she took the leap.[12] She and Jarrah became a couple. They cooked meals together. She helped him learn German. Bleary-eyed photographs from the time, including one of Jarrah lighting a water pipe, indicate that they did their share of partying.

Not everybody joined in. One man Jarrah later grew close to, Abdulrachman al-Makhadi, one of Aysel's dental school classmates, was known around campus as the self-appointed enforcer of Muslim doctrine; he governed harshly from inside a small cinderblock mosque that locals referred to as "The Box." Inside it, Makhadi, a Yemeni, preached a strict interpretation of Islam and collected money for the Palestinian militant group Hamas. Outside, he didn't hesitate to badger other Muslims into attending prayers. Jarrah went only occasionally, usually on Fridays, and more as a social than a religious occasion.[13] Jarrah and his friends would go to prayers, then try to escape for beers afterward, but Makhadi didn't let anybody leave. He would stay and preach after prayers until two or three in the morning.

"I told [Makhadi] this is not right," said Assem Jarrah, Ziad's cousin, who had lived in Greifswald when Makhadi started the mosque. "God doesn't need this." It didn't stop Makhadi, however. Among other things, the little mosque provided a livelihood for the Yemeni, who was married and had several small children. He purchased Arab food in Hamburg, brought it back to Greifswald, and resold it at the mosque, Assem said.

Jarrah went home for the winter holiday after his first semester and upon his return seemed changed, no longer the happy-go-lucky playboy. His cousin, Salim, noticed that he began reading a radical Islamist publication called *Al Jihad*.[14] A friend of Aysel's said that in early 1997 Jarrah had talked to her about being "dissatisfied with his life up till now." The friend said Jarrah wanted to make a mark in life and "didn't want to leave Earth in a natural way."[15]

There was no indication of what lay behind the change.[16] That year, back in the Bekka Valley, a new Sunni Islamist movement had begun forming, using the Internet and small group meetings to spread the word. It was growing rapidly when Jarrah was home over break, but there is no evidence he came into contact with it. Salim Jarrah said once that his cousin was like a tree without roots; Greifswald was no place for an Arab to grow them. The city might as well have been in a different universe than sunny, cosmopolitan Beirut, which, like many capitals, regards itself as the center of the known world. Stylish, flashy Beirutis rival New Yorkers and Parisians in their provincial conceits. Salim and Ziad laughed when they first saw what passed for a place to party, the Fly-In Disco, in Greifswald. The town was a dim, almost medieval place that seemed decades behind the rest of Germany, not to mention Beirut. Fashions in clothing, even today, seem stuck twenty years back, and the city has for years had a large population of neo-Nazi skinheads. It was not a welcoming place for foreigners.

Jarrah's new piety caused problems with Aysel almost at once. He criticized her choice of friends, the way she dressed, and what she drank. The couple were in many ways dissimilar. He was often quiet and withdrawn. She talked all the time about everything to whomever would listen. "That's my way. That's how I am. I tackle problems through conversations," she said. Throughout the relationship, according to their correspondence, she railed at Jarrah for not telling her more, for not sharing more of himself.[17] Jarrah habitually responded that he told her what he felt she needed to know.

At some level, she must have understood what Jarrah was going through. She had earlier experienced a crisis of her own. After high school, her parents had sent her to Turkey, apparently an attempt to ground her in her heritage. It backfired. She attempted suicide. "I was in a cultural conflict," she said.

"When he asked me to change, I sometimes said, 'OK, you are right,' but I didn't do anything. I know that kind of culture—that's not so different with Turks."

One thing they agreed on 100 percent was getting out of Greifswald. Upon his arrival Jarrah had enrolled, as foreign students are required to do, in preparatory German classes. He was due to complete those within the year, and in the spring of 1997 he began to apply for regular university admission. He wanted to study dentistry like Aysel and he applied at medical schools around the country. Later, out of the blue, he also applied to the biochemistry program at Greifswald, probably as a fallback position, and to study aeronautical engineering in Hamburg. He told Aysel Hamburg was the only place he was accepted.[18] This was untrue. He was accepted into a medical school in western Germany and in the science program in Greifswald and at Hamburg, which is the one he chose. He offered no reason for the choice and had no obvious connections to the city, but a good friend, Bashir Musleh, had met and befriended several students from Hamburg at a summer job the previous year. Bashir was going to Hamburg; plus, Makhadi knew many of the men at Al Quds Mosque, in Hamburg, and was moving temporarily to Hamburg for an internship.

Jarrah moved to Hamburg with Musleh and enrolled at the University of Applied Sciences, which was the university preferred by the young Moroccans from Al Quds. The two men moved into a student apartment house with Abbas Tahir, an older Sudanese student who was already fully immersed in the broad circles of fundamentalist Islam throughout Germany. Tahir, over time, became Jarrah's confidant, the man he called when faced with important decisions. In Greifswald, there had essentially been one radical Islamist—Makhadi. In Hamburg, there were dozens, maybe hundreds. Through Tahir, Jarrah would meet many of them. In the beginning, Jarrah returned to Greifswald every other weekend to see Aysel. He sometimes rode the train with Makhadi, whose family had stayed behind in Greifswald.

RAS AL-KHAIMAH

Another Hamburg newcomer was a young man named Marwan al-Shehhi, a native of the United Arab Emirates who had come to Bonn, Germany, as a prospective marine engineering student in 1996. He was on scholarship from the UAE Army, not the world's most effective fighting force but one of its most generous, paying the students monthly stipends of about $2,200 apiece.

Shahid Nickels said Shehhi was extremely laid-back, "dreamy, lumbering, slow, docile, slightly spoiled, an easygoing bon vivant and romantic. He was friendly, always in a good mood, well-educated, humorous, and sometimes a little clumsy. . . . He also had this certain sense of irony, saying stuff never directly, but just hinting it verbally or with his gestures. He never spoke negatively about others and never used a negative word. He never looked stressed. . . . He radiated a sense of calm."[19]

Shehhi was a rotund man and—the opposite of Amir—he loved to eat. He always took twice as long at the table as everyone else and enjoyed every minute of it, unself-consciously rolling rice into little balls between his thick fingers and humming jihad songs as he chewed. He had an incurable sweet tooth and often carried a bag of candy, which he shared readily with everyone. This was as far from Amir's "reluctance to all pleasure" as one could get, although candy—chocolate, in his case—was Amir's great vice, also. With his monthly UAE Army stipend, Shehhi was relatively better off than his friends, and sometimes it showed. He was nearsighted and bought expensive eyeglasses—one pair cost nearly $400. He occasionally bought expensive clothing, too, and a few times rented big German luxury sedans, but he wasn't wealthy and his sartorial tastes tended toward a mismatched pair of suit pants with tattersall-checked shirts. In fact, his family was humble and when people automatically assumed he was rich just because of where he was from, he resented it. This happened often; some of the hard-core Moroccans used to invite him to dinner in the hope that he would pay for everyone.[20]

Shehhi's German was not good, but his Arabic was superb and his knowledge of Arab literature surpassed everyone else's. Although the youngest of the core group, he was the acknowledged authority on Islamic scriptures. He occasionally entertained everyone with comic recitations of Arabic fairy tales. In a culture where talking was prized, Shehhi shone.

Shehhi never voiced the slightest doubts about his beliefs. He never spoke about women as anything other than potential marriage partners and never spoke *to* them at all, unless compelled. There was nothing the least bit secular in his background. As one of the few Gulf Arabs in the Harburg group, and the son of a religiously trained father, Shehhi had more formal Islamic education, and had lived a stricter version of Islam his entire life, than the others.

The Emirates is a federation of seven sheikdoms on the southern end of the Persian Gulf. It is known to the rest of the world, to the extent that it is known at all, for the glitz and glamour of Dubai, a city that has grown into a combination of financial capital and party town; with all the bright lights, outlandish architecture, and women of questionable reputation of a Las Vegas, plus lots of big banks. Dubai, however, is just one of the emirates and markedly different, more urban, Western, and cosmopolitan, than the others. Shehhi is from the emirate of Ras al-Khaimah, 70 miles and one or two centuries northeast of big-city Dubai.

Almost all the wealth of the Emirates is from oil and gas and almost none of those reserves belong to Ras al-Khaimah, which is at best a poor cousin and looks it. A goat market on one edge of town is among its most visible enterprises. There's not a whole lot between the goats and the other end of town, which is where the Shehhi family home was located. Goats graze through the neighborhood there, too, although the grazing is sparse among the rocks and the sand, so much so the goats survive mainly by eating away at the lower branches of a few ragged trees.

The Shehhi home is modest, a small, dowdy cinderblock construction with heavy metal doors; its white and pink paint has long since paled under the force of desert sun and wind-whipped sand. Shehhi's half-brother, a retired policeman, lives in a much larger, newer, and more lively looking house across the dirt road. Shehhi's father was the local muezzin, the man who called people to prayer at the neighborhood mosque, which abuts the home. Shehhi was close to his father and made the prayer call when the father could not, locals said.

Shehhi joined the UAE Army not long out of high school.[21] After a few months' initial training, the army rewarded him and three fellow recruits with college scholarships to Germany, although it was not clear from his record that Shehhi had any serious scholastic ambitions. He and the other young men arrived in Bonn in the spring of 1996. Shehhi took a course at a German language institute, then was admitted to the next year's preparatory studies at the *kolleg* in Bonn. He went to classes and studied diligently, but wasn't very successful. His father died in the spring of 1997, and Shehhi requested a leave from school. The army denied the leave, but Shehhi left anyway and subsequently failed his course that year. He returned and retook it the next year and passed.

Shehhi went to Friday prayers every week with other expatriate emiratis at the UAE embassy in Bonn, and otherwise tried to abide by the tenets of Islam. Some of his countrymen couldn't understand this. They came to the West and intended to act like Westerners, they said.[22] And that included dating women and drinking alcohol, two of the more notable taboos in Islam. His friends couldn't understand Shehhi's reluctance to join in the fun. Some of them resented it,[23] but Shehhi was resolute, even insisting they not eat at restaurants that served alcohol, not an easy decision in Germany. The young Arabs almost never cooked, and probably didn't know how. Shehhi's strictness made eating out much more complicated, but he insisted and the others grudgingly abided by his dictum. They ended up in grimy fourth-class Turkish fast-food cafés, just about the only places in Germany where beer was not served. Rather than relaxing his attitudes as he settled in to German life, Shehhi became stricter. After his father died, he refused even to go into a McDonald's restaurant because he had heard they used pork fat to make their french fries.[24] He lived with a German host family and apart from being irritated when the woman of the household asked him why he had no girlfriends and never drank alcohol, he got along passably with them, although they said he was a fairly cold presence in the house.

Then that winter, out of the blue, he asked his embassy if he could transfer to the *kolleg* in Hamburg.[25] The embassy military attaché made inquiries on his behalf, and he moved to Hamburg early in 1998. He was immediately accepted within the Harburg group, and his personality brightened considerably. The scowling, dour ascetic in Bonn was transformed into an easygoing, robust, hail-fellow-well-met young man who loved to sing and laugh. (This was true of Omar, too. People who knew him well described him as having an exceptionally sunny disposition; those who knew him less were apt to describe him as hectoring and unpleasant. Maybe Amir was so dour all by himself that everyone else just seemed lighter in comparison. Or, more likely, they felt comfortable and at ease with one another in ways they never did with outsiders.)

AFGHANISTAN

Amir left Hamburg as he usually did over the winter holiday at the end of 1997. This time, he didn't return for three months. When he finally came back, he told his roommate he had been on another pilgrimage to Mecca.[26] He had been on *hajj* just 18 months earlier, and it would be unlikely for a student—even one so devout—to go twice so quickly or stay so long. It was the longest break in his schedule since he had come to Hamburg and there is no record that he spent any substantial portion of it at home in Cairo. It seems likely he went to Afghanistan to the jihad camps. Among other indicators, the most telling was that after he returned in the spring he applied for and received a new passport, although his old one had not yet expired.[27] He said he had lost the old one. This was common practice among jihadis, done in order to erase visible evidence of their travel. Whereever he had gone, his friends held a party for him in the spring, not typically what they did when somebody went home for a while.[28]

Omar and Shehhi were also gone from Hamburg for long stretches later in 1998. Omar was gone for much of the summer, then again in the early winter.[29] Shehhi was unaccounted for during most of the fall and early winter. Before he left, he withdrew more than $5,000 from his bank. While he was gone, his normally active credit card accounts were dormant. He made no charges on them or cash withdrawals from automated teller machines from September 3 to early December. As with Amir, there is no documentary evidence that they went to Afghanistan, but there is also no indication of where else they might have been. People don't usually disappear so completely unless they intend to. Practically, there is only one place they likely would have gone—Afghanistan.

Going to the camps was not all that unusual among young Arab men. Tens of thousands from across the world moved through the camps. Most of these men returned to wherever they lived not much changed. The camps offered a huge variety of experience and, more importantly, a huge variation of intensity of experience. It mattered most who the man was who went in order to understand who the man was who came home. For many, the camps were little more than a lark, an adventure vacation. For the Hamburg men, vacation did not seem to be what they were looking for or what they found.

WILHELMSBURG

After Amir returned to Germany in the spring of 1998, the group for the first time began to act as a group, not as a collection of individuals. Almost everything the core members did, they did with the others. Omar left the container camp for good and lived with one friend or another. In summer, Essabar, Motassadeq, and Jarrah, worked in the paint shop of the Volkswagen plant in Wolfsburg. Amir, Omar, Shehhi, and Belfas, among others, all worked in a warehouse together, packing crates of computers for shipping. The man who owned the company said he routinely hired students when he had extra work; this is how he had come upon the group. He found them by calling people who had worked for him previously; they called others. It was normal summer work for students. But Belfas? Even the man who owned the company thought it odd that a middle-aged night postal worker would spend his days in a computer warehouse.[30]

Amir had finally exhausted his eligibility for university housing. He had been in his student apartment for nearly six years, a year beyond what was normally permitted. He told his house manager, Manfred Schröder, that he would take a nearby apartment with friends.[31] Schröder was probably the only one sorry to see him leave. He was an older man, tall and erect with an air of almost military authority. Not all the students appreciated it. Amir, though, habitually treated older men with deference. He sometimes invited Schröder into his apartment for tea and chocolate. Schröder had granted Amir one extension on his housing already, but he had been at Centrumshaus too long. Now Schröder told him he had to move.

Amir left the student house that summer. He and a group of men— nobody knew quite how many—moved into a project flat in Wilhelmsburg, an island in the middle of the Elbe River, at a red-brick prewar housing development on a broad, bleak street that faced a ribbon-wire fence and the forbidding prospect of the Hamburg harbor beyond. Wilhelmsburg was a worn-out industrial suburb, so psychologically remote from the rest of Hamburg it was sometimes called the Forgotten Island. People passed over it on commuter trains or autobahn bridges without even realizing what was outside the window. It was there but hidden. If your aim was to vanish, to drop from the face of the world and yet keep the world close at hand, this was a good place to go.

They rented a third-floor, two-room walk-up for $250 a month in one of several side-by-side six-story buildings. The other flats were filled mainly with working-class Turks, by far the largest minority group in Germany, and a fair number of pensioners, drug dealers, and prostitutes.

The group had gone to ground. There was almost a feral quality to their activities. They were seldom seen outside and when they were they seemed to be moving in a pack.[32] Inside, they kept the blinds drawn shut night and day. They had no furniture, only mattresses. They piled their clothes in a corner of the living room floor. They spread newspapers on the floor in place of a tablecloth and ate their meals sitting there. There was no telephone or, apart from the lights, electrical devices of any kind. The men talked long into the night most nights and disappeared all day most days, said Helga Link, a downstairs neighbor. Link lived directly beneath their apartment and could hear every footfall on the hardwood floors. She never once heard a radio or television or a single note of music—just soft stockinged footsteps and the voices of men talking.

The first address Shehhi registered upon his arrival in Hamburg was the Wilhelmsburg walk-up. He very quickly became an integral—and in many ways welcome—addition to the Omar-Amir group.[33] For one thing, he was a happy presence—a lightener of dark moods. For another, he seemed content to serve as Amir and Omar's loyal acolyte, never questioning or challenging them, but lending them the authority of one who knew, not just one who had learned. He had a deeper background in matters of faith and carried a sense of moral authority, but did so lightly.

Amir organized the group, enforced the rules. Or at least tried to. Once, when he, Omar, Maklat, and Musleh were walking back to the station to catch a train after prayers at Al Quds, Amir decided they should walk on a different street in order to avoid the prostitutes on Steindamm.[34] This struck the others as silly because the mosque and the train station were on the same street and they walked it virtually every day. Musleh told Amir to close his eyes if he couldn't bear it. Amir insisted on the detour. The others refused, and Amir, ever stubborn, took it by himself. When Amir was not around, the other men mocked his sternness and complained about all his rules. Amir had also begun to use a sort of rouge makeup; no one knew why. People made fun of him for that, too.

Omar was the one who connected the group to the wider world. He seemed to have contact with everyone. One day he was arranging to get a satellite telephone or to meet some Kosovo Albanians he knew, the next he was off to Munich or Greifswald. He always had a cell phone but changed the account so often it was hard to keep track of what his current number was. He kept the phone along with a pen and notepaper in the pocket of the vest he always wore. How he paid for the phones was unclear. Once he escaped the asylum camps, he never held a real job. He often slept in until near noon, then began a constant circuit of mosques, student homes, and Islamic study groups in Hamburg, often visiting as many as ten residences in a day. He was habitually late for everything, too slow to leave the last place to get to the next one on time.

In the course of his visitations, he met Ziad Jarrah not long after Jarrah had arrived in Hamburg and drew him into the outer edges of the group. He traveled extensively within Germany and could easily have met Shehhi on one of his trips. He was often sly about his travels, answering questions with conspiratorial grins. Unlike Amir, who would stare stonily at the floor in the presence of women, Omar stole glances at girls on the street. If caught looking, he'd smile his shy smile and look away.[35] He talked about women often, although never when Amir was around. He sometimes told friends he was looking for a wife, and in the course of his travels, he actually did develop a serious relationship with a woman in Berlin and stayed with her when he was in town.[36]

Omar and Amir regularly lectured and recruited for their own study groups in the mosques, although they were better at finding people than keeping them. Amir, in particular, had such a stern visage and uncompromising vision of the practice of Islam that he drove students away. Shahid

Nickels said at the end, when the classes had run their course, Amir "was sitting there almost alone."

Amir's study groups were rigidly structured and roles within them assigned: A passage of the Qur'an was read (by someone of Amir's choosing), followed by an explanation of the passage and the comments of the Prophet on it, which would themselves in turn be explained. Then Amir would speak on the subject addressed by the readings. Only after all of this were questions allowed. If anyone attempted to deviate from the established routine, Amir would grow visibly upset, chewing on his lower lip in agitation.[37] When Maklat sought a formal role as a reader in one of the study groups, he was denied. "You are too weak," Amir said. Once, when Maklat asked why Amir never laughed, Amir said: "How can you laugh when people are dying in Palestine?"[38]

"Joy kills the heart," Amir told him on another occasion. Even Belfas, who regarded Amir almost as a son, told him once: "Maybe you are, with all your strictness, sent by the [Jews] to cause confusion among the Muslims."[39] Amir told friends he was losing respect for Belfas because he had grown soft.[40]

One of the few times anyone could recall Amir ever laughing at anything was when a group of the men watched a television news documentary on the Intifada at Belfas's apartment. Volker Harun Bruhn, who was studying Islam under Belfas, said the program told the story of a Palestinian suicide bomber, who set off his charge prematurely, injuring only himself.[41] He was rushed to an Israeli hospital unconscious. He awoke on the operating table, looked up at all the doctors and nurses in white clothing and masks gathered around him and asked: "Is this heaven?"

A doctor replied with a question of his own. He asked the bomber if he thought there would be Jews in heaven.

The bomber replied, "No."

"Then," the doctor said, "I guess you're not there yet."

This cracked everybody up, Bruhn said, even Amir.

Despite Amir's sternness, a spirit of easy brotherhood often prevailed within the group. However extreme its aims, it was a kindred community. The men shared apartments, bank accounts, and cars. The group members strictly observed the tenets of their religion, but they laughed about how inept any of them would be if they had to actually butcher a lamb according to religious custom for a holiday feast. They prayed five times a day, often holding hands when they did. They maintained strict Islamic diets and even debated the proper length of their beards. Shehhi was the beard expert and declared the appropriate length was long enough that a fist clenched around it would leave at least a couple of inches of beard exposed below the bottom of the hand. They talked endlessly about conspiracy theories and the damage done by Jews, including their unquestioned assumption that Monica Lewinsky was a Mossad agent sent expressly to bring down President Clinton. For entertainment, they watched battlefield videos from Chechnya and sang songs about martyrdom. Shehhi loved to march through their apartments while he and Omar sang jihad songs. When Nickels showed up, they would greet him, laughing and singing the song, "Ya, Shahid (Oh, Martyr)."

During Ramadan, the men made a tour of Hamburg's Muslim community, breaking their fasts with Indians, Malays, and Turks, whoever would have them. Even then, Amir would attempt to make their hosts adhere to his notion of what rules to apply. Once, at a Turkish home, he even went into the kitchen to hurry along the meal because the proper timetable wasn't being followed. Habitually, Omar followed behind and played the peacemaker, joking and soothing hard feelings.[42]

Omar was dreamily romantic about the jihad. "It is the highest thing to do, to die for the jihad," he told people. "The mujahideen die peacefully. They die with a smile on their lips, their dead bodies are soft, while the bodies of the killed infidels are stiff."[43]

Everything in Hamburg took on a new urgency. Living arrangements became increasingly fluid. Nickels said that the group members always looked as if they were ready to leave wherever they were at a moment's notice. The men moved in and out of flats. They began physical fitness training, although their intentions often outweighed their performance. Sometimes, a workout consisted of nothing more strenuous than a walk around the block. One of them, Said Bahaji, joined the German Army, then sought a medical discharge after completing combat training. He could have avoided service altogether (he was asthmatic), and initially opposed joining, but he suddenly changed his mind, apparently wanting the training. Shehhi, too, made a feeble attempt at conditioning. He bought a bicycle and pedaled it around Harburg, preparing for jihad, he said.[44] The men tidied up personal affairs, assigning powers of attorney and control of bank accounts to friends. They rushed to finish school courses or gave up all pretense of trying.

The pace had quickened in the broader jihad community, too. In the spring of 1998, Osama Bin Laden had issued his second and most public fatwa against the United States, a call for a crusade against the Americans and their surrogates around the globe. A battle horn had sounded. The event gathered little notice in the United States, but it was instrumental in the continued strengthening of Bin Laden's image among potential jihadis. Bahaji, for one, was an avid follower of Bin Laden's career. He downloaded Bin Laden speeches and writings to his computer and made dozens of copies of those he liked.[45] Bin Laden was seen as a man who would stand up to the United States and would do so with more than mere words. The war was on. In the summer, two American embassies in East Africa were attacked with truck bombs, and hundreds were slaughtered. That most of the dead were African Muslims seemed not to matter to true believers.

HARBURG

The group didn't stay long in Wilhelmsburg. That winter, Amir, Omar, and, after he returned to town, Shehhi moved into a clean, neat, newly refurbished apartment at Marienstrasse 54, an apartment house on a sloping, treeless street just a five-minute walk from TUHH, where Amir and Shehhi were still enrolled. The apartment was a huge improvement over the Wilhelmsburg flat: it had three bedrooms, a full kitchen, new paint, and a new furnace. It was like moving from the third world to the first. Some friends referred to the flat as Dar al Ansar, the House of the Followers, and noted it as such in their phone books.[46]

They were good renters, said Thorsten Albrecht, their landlord. They paid the rent on time and didn't cause trouble. Albrecht thought they looked and acted like philosophy students. They always seemed pensive, preoccupied. They dressed either in Arab tunics or in Western clothes that had been out of style for a while. Albrecht, a typically stylish Hamburg resident, remembered with disdain one of them frequently wearing beige bell-bottom jeans.[47]

The tenants at Marienstrasse were almost fungible. Over two years, more than a dozen men listed it as their residence. They moved in and out as their needs or inclinations changed. Shehhi moved out after only a month, taking an apartment by himself nearby. He said he intended to return to the UAE, find a wife, marry her, and bring her with him back to Harburg. Bahaji, who as a German citizen had signed the original lease, replaced Shehhi.

Bahaji was an electrical engineering student at TUHH and an able computer programmer, according to his bosses at a company where he worked part-time. In many ways, he was a typical computer nerd—an autodidact, as one man described him—who seemed more at ease at a keyboard than in social situations.[48] When he had come to Hamburg at the beginning of 1995, it was a homecoming of sorts. His Moroccan father and German mother had met and married in Germany, and Bahaji was born there in 1975. The family moved to Morocco when he was 9 years old. He returned to Germany for university.

He was in some ways a typical German undergraduate, a fan of video games and Formula One autoracing.[49] In his first year in school, he met and fell in love with a German Catholic girl who had grown up in Brazil, and so was, like him, neither entirely an insider nor an outsider in the student culture. When the relationship grew serious, the girl's parents ordered an end to it, sending her abroad to ensure its demise. Bahaji, during that period, lived in a student dormitory and spent weekends with his aunt, Barbara Arens.[50] Arens said he was heartbroken and embittered by the end of the love affair. He also resented the fact that he didn't have as much money as many of the students, and so didn't socialize much with them. He was additionally stigmatized by speaking with a slight lisp and walking with a limp. It wasn't long before he found a new circle of friends among Moroccan students, including some who were regulars at Al Quds. His aunt said he began to change markedly and rapidly. Within mere months, Arens said, the "pro-American, pro-German, pro-European" student was lecturing everyone he knew about Islam. This included the aunt, who wanted none of it. They drifted apart, and eventually she refused to let him stay at her home.

Mounir el-Motassadeq, a fellow Moroccan, met Bahaji at Al Quds and introduced him to Amir and Omar. Even though Bahaji had almost no formal Islamic education—the other members of the group had to teach him how to pray—he quickly became one of them.[51] By the time he moved in with Amir and Omar, he wrote to his old girlfriend, he no longer grew angry when Germans called him a foreigner because he didn't "feel German at all anymore."[52]

As time went on, the focus on religion became almost an obsession within the group. Discussions intensified, although, friends said later, they were scattershot. One week, the members were intent on fighting in Kosovo, the next in Chechnya or Afghanistan or Bosnia. The men were agreed: they wanted to fight—they just didn't know which war.

While Amir and Omar sought to instruct, some of the younger Moroccans from Al Quds, many of whom worked together at the Globetrotter outdoor supply shop, were more forceful. They appointed themselves as enforcers, a sort of informal religious police, who took upon themselves the duty of insuring that other young Arabs in the city were sufficiently devout.[53] They once beat up a fellow employee because he refused to grow a beard, and threatened other students at knifepoint.[54] Another time, Motassadeq proclaimed that any woman who wore a skirt that was too short deserved to be raped.[55]

They did not allow friendship to temper their demands, as Essabar illustrated in his relationship with Yassir Boughlal. Essabar and Boughlal had met not long after both arrived in Germany from Morocco, and they studied together for a year in Koethen.[56] In Morocco, Essabar had attended a military academy and was slated to become an army physician. It was a prestigious path that he suddenly abandoned just before he went abroad to Germany. Boughlal said it was the one thing Essabar never talked about. In Koethen, they became fast friends. They met each other's families. Essabar was the best man at Boughlal's wedding. Boughlal had always been the more secular of the two; he smoked and drank beer, and his new wife wasn't even Muslim, but the men had enjoyed one another's company and shared confidences. Then they moved to Hamburg together and something happened. They both enrolled at the University for Applied Sciences, but otherwise drifted apart as they made other friends. Essabar had begun attending Al Quds, and within months he changed dramatically. "Suddenly he was so extreme, so stern about religion. He stopped joking. He never explained to me why he changed," Boughlal said.[57]

One day, after having not seen Essabar for a while, Boughlal received a phone call from him at his apartment.

Essabar asked: "When will your wife convert to Islam?"

"Never," Boughlal answered.

Essabar called back again and again, asking the same question. If Boughlal's wife accidentally came on the line, Essabar hung up. Then, as suddenly as he had begun the calls, he stopped; the two men didn't speak for months. Then Essabar called one more time. He asked: "What about your wife? Did she convert?"

"No," Boughlal said. They talked briefly, then Essabar said goodbye. "Where can I reach you?" Boughlal asked.

Essabar replied: "You can't reach me anymore."

This was, in almost every sense, true. The Hamburg men who joined their plights to that of fundamentalist Islam chose not simply a new mosque or religious doctrine but an entry to a new way of life, the acquisition of a new world view, in fact, of a new world. And, as with Essabar, it often didn't take long. Essabar and Boughlal had known one another for more than a year, and after Essabar "converted" and Boughlal did not, that was the end of their relationship. Essabar even told his own father to grow a beard. Boughlal's father visited him once from Morocco and went with his son to Al Quds for prayers. Boughlal said it was a routine day at the mosque, but his father was so unnerved he told Boughlal never to set foot in the place again lest he be imprisoned upon his return to Morocco, which, like most Arab countries, regarded radical Islamists as threats to the stability of the state.[58]

At the core of the group's emerging beliefs was their sense of obligation to perform their duties for Allah. "Omar thought of the jihad as the big challenge," Nickels said.[59] "It is a duty for the Muslim society to always have fighters for the jihad. The Muslim society is like a body: if one part is hurt, the whole body suffers. He said a couple of times that it is the highest deed for a Muslim to die in jihad."

One day, in a grocery store, Amir asked Maklat: "Are you ready to fight for your faith?"[60]

"Not yet," Maklat said.

Amir answered: "The brothers die in Bosnia-Herzegovina and you say no."

Another time, Amir criticized Maklat for buying expensive food: "You are living like you're in a paradise, while brothers die elsewhere."[61]

Omar emerged as the emotional leader of the small group. Amir set policies and decided what the group should study and how, but Omar was the one who gave them their sense of purpose.[62] He told the other men they were obliged to go to those places where Islam was under attack, to defend it literally as holy warriors. He played cassette tapes of jihad songs and battlefield propaganda videos about Chechnya, Bosnia, and Kosovo. He showed the videos all over town, in student dormitories and private apartments. One of the key tenets Amir and Omar preached was that no matter where they fought, their real enemies were the Jews, and ultimately the Americans. "One has to do something about America," Omar said.

Once, when Nickels complained that the plight of the Palestinians was hopeless, that Israel and America were too powerful and nothing could be achieved by confronting them, Amir said: "No, one can do something. There are ways. The U.S. is not almighty."[63]

Although Amir and Omar seemed to agree on many things, they had different emphases. Omar was motivated more by religious belief, and Amir by politics. "With Amir, it was always hate when it came to the Jews. He was very emotional about the political issues, while Omar was emotional about the religion," Nickels said. Amir saw a worldwide conspiracy at work, bolstered by the Americans, but run always by Jews. He blamed Jews for almost every wrong imaginable. Once, when he and Nickels were alone in the flat, Amir went to the bathroom to relieve himself. In the course of it, he made so much noise Nickels heard him from the living room and couldn't help but laugh. When

Amir emerged, he blamed the Jews for having built the bathroom's too-thin door.[64]

In his discussion groups, Amir frequently talked about the divisions within Islam, how there was no united Muslim nation.[65] Omar was an admirer of Osama Bin Laden, referring to him as a great man, while Amir withheld judgment. Maybe he is, he said. And maybe not.[66]

CHAPTER 5

The Smell of
Paradise Rising

HARBURG

THE FLAT AT MARIENSTRASSE 54 became the hub of increasingly excited activity. The apartment became a kind of clubhouse for young Hamburg jihadis who would gather for an evening meal and conversation. Amir or Bahaji cooked simple foods—chicken and potatoes or plain rice and bread. Neighbors would sometimes count more than a dozen pairs of shoes lined up outside the door. Inside, the flat was sparsely furnished—a step up from the down-at-its-heels summer camp atmosphere of the place in Wilhemsburg. A few religious posters were hung in the hallway, and that was it for decoration. There was a table and chairs in the kitchen and two of the three bedrooms had actual beds. The third, Omar's room, had as usual only a mattress on the floor. Amir and Bahaji also had computers in their rooms and textbooks mixed in with the Islamic works. Omar had only religious books and cascading piles of cassette tapes and videos, all having to do with Islam and jihad.

As it turned out, the neighbors weren't the only ones counting the shoes in the hallway. By 1998, the radical Islamist scene in Hamburg had reached a level of activity authorities could no longer ignore. In Germany, for most of the 1990s, the threat from any sort of internal dissent had seemed distant. The last real concern had come from the political left in the form of the Red Army Faction, a successor to the 1970s Baader-Meinhof gang. But that threat had been ended years before. The class struggle was history. The only thing the young Germans of Generation Golf shared with the Baader-Meinhof Maoists was an affinity for black turtlenecks. One measure of the seriousness with which Hamburg viewed the potential threat from radicals within its fast-growing Muslim population was the distribution of counterterrorism resources. Local police had one man assigned part-time to monitor radical Islam. That's half a man to look for dangerous activities among a population of 80,000 people.

The young men at Marienstrasse 54 were not at the center of anyone's attention; they had done little to attract much beyond curious stares and whispered gossip from their neighbors. But the paths they cut through Hamburg's Muslim underground had crossed those of people who *were* the focus of official scrutiny. And when the paths began to intersect so often that they seemed to intertwine, the Marienstrasse men inevitably became subjects of interest themselves.

The scrutiny of the group focused not on its leaders—Amir and Omar—but on others less central to it. The first link to them came through a Syrian native who had been living in Germany for sixteen years, married a German woman, and become a citizen. His name was Mamoun Darkazanli.[1] He had been a part of the first wave of Islamist activists to settle in Germany after fleeing Syria in 1982. In Hamburg, he had established a small trading company; he lived a quiet life and did not arouse suspicion until the middle 1990s. He first came to official notice when his telephone number was found in the possession of a passport forger arrested in Africa in 1993.[2] A brief investigation failed to uncover any evidence of wrongdoing by Darkazanli. Then, after the bombings of the American embassies in Kenya and Tanzania in February 1998, a Lebanese-American suspected of playing a role in the bombings was arrested in Kenya. He had in his possession business cards for a trading company he claimed to operate. The man lived in the United States, but the address for the company was in Hamburg. It was Darkazanli's home address, an apartment house in a sensible, pleasantly leafy middle-class neighborhood. This renewed interest in Darkazanli; closer inspection revealed that the trading company didn't seem to do much trading, and it wasn't immediately apparent to anyone how Darkazanli earned a living.

That fall, Mamdouh Salim, an Iraqi jihadist suspected of involvement in the embassy bombings, flew to Bavaria. Salim had been an associate of Osama Bin Laden since the days of the Soviet war. When Bin Laden was expelled from Saudi Arabia after the war and moved to Sudan, he gathered many of his former "Afghan Arab" colleagues with him. Among them was Salim, who became an executive in two Bin Laden companies and also sat on the governing council, or *shura*, of the fledgling Al Qaeda organization. He was regarded by American intelligence officers as a key Al Qaeda operative, one tasked with, among many other duties, attempting to procure uranium for Al Qaeda's nuclear weapons efforts. When the Americans discovered that Salim was scheduled to fly from the United Arab Emirates to Germany, they asked the Germans to arrest him.

Salim had traveled extensively throughout the 1990s on behalf of Bin Laden and had been to Germany at least four times previously. The Germans acquiesced to the American request, which was accompanied by a warrant, and arrested him. Under interrogation, Salim admitted his business associations with Bin Laden. He protested, however, that he knew nothing of Bin Laden's alleged terrorist activities and said he had come to Bavaria—albeit in a roundabout way, via Turkey and Majorca—merely to buy a used Mercedes station wagon. He offered no convincing explanation as to why he would choose the obscure small town in which he was arrested to make the purchase. Plus, he carried only $800 with him, hardly enough to buy a set of tires, much less an entire Mercedes. Investigators discovered Salim had Darkazanli's telephone number programmed into his cell phone and subsequently determined that Darkazanli had power of attorney over a bank account Salim had established years earlier in Hamburg.[3]

The investigation eventually led German security services to seek a criminal indictment of Darkazanli, but prosecutors rejected the request, saying the evidence was insufficient. The evidence showed a man with close acquaintances among fundamentalist Muslims, including some associated with Osama Bin Laden, both in Germany and abroad. This was not, as a German investigator noted, illegal.[4] It was of sufficient concern to put Darkazanli under limited surveillance, though. Among the people revealed to be acquainted with Darkazanli was one of the tenants at Marienstrasse 54: Said Bahaji.

Bahaji, although he had been one of the last to be assimilated into the Amir-Omar group—and at the beginning the most Western of them all: Christian girlfriend, pop culture aficionado—had quickly become one of the most fundamentalist, given to making especially harsh statements against nonbelievers.[5] Once he had started on this new path, he seemed intent on speeding along it. He constantly surfed the Internet for Islamist sites[6] and became a devotee of Osama Bin Laden. His transformation from an auto race fan who didn't know how to pray in Arabic to aggressive would-be jihadi was startling and complete. At the Al Quds mosque, investigators noticed that he met frequently with Darkazanli and also with another Syrian immigrant, a friend of Darkazanli's named Mohammed Haydar Zammar.

Like Darkazanli, Zammar had been settled in Hamburg for years. He was a big, beefy, heavily bearded, middle-aged, unemployed auto body mechanic who, along with his wife and six children, lived on state welfare payments. No one could remember the last time he had held a job. He spent so much time proselytizing for Islam it was hard to imagine when he might have found time to work. Zammar's bluster matched his 6-foot-4, 300-pound physique. In almost any discussion, his was the loudest voice and most radical view. He was well-known in many of the city's mosques as an advocate of jihad. He had trained in the Arab camps in Afghanistan in the early 1990s, had been in Bosnia during the war there, and regularly urged younger men to follow his example.

Not everyone felt that Zammar should be taken seriously. The president of one mosque where Zammar was a frequent visitor called him merely "a little boy in a big body" who talked too much.[7] Zammar's own brother said he was harmless, but that "his tongue was his problem."[8] He could seem downright ridiculous at times. Once, in talking about the Afghan training camps, Brother Haydar, as he was known, bragged about having learned self-defense there. He illustrated with a sideways judo kick and fell flat on his face, shaking the building so much the downstairs neighbors came to see what had happened. But for all of Zammar's buffoonery, he was well-connected in the Islamist underground, and German intelligence had already opened an independent investigation on him.

That began at least as early as 1997, when Turkish intelligence agents informed their German counterparts that Zammar was running what amounted to a travel agency for jihadis, routing them through Istanbul and Ankara en route to Pakistan and Afghanistan.[9] Zammar himself was said to have made several dozen trips through Turkey to Pakistan and, presumably, Afghanistan. He was known to be an avid Bin Laden admirer and acquaintance. The Germans initiated surveillance of him, including phone taps and physical observation. They gave the investigation the ironic code name Operation Tenderness. (Zammar's name begins with the same two letters for the German word for tenderness, *zartheit*.) The surveillances of Darkazanli and Zammar led authorities to Bahaji and Mounir el-Motassadeq, and from them to the Harburg apartment. Once you knew any of the Harburg men, it was hard not to know all the rest.

The Germans thought enough of what they learned to pass some of the material and names on to their American intelligence counterparts. The Americans already had a longstanding interest in Zammar, whom they knew to have connections to a wide range of Islamists.[10] Among other things, the Germans told the Americans they were tapping Zammar's telephone. In the course of that surveillance, they recorded several conversations between Zammar and Shehhi. The first of these occurred while Shehhi was in Bonn. The Germans told the Americans Zammar had spoken with a man in Bonn named Marwan, who they suspected might be connected to Al Qaeda. They gave the Americans Shehhi's UAE mobile phone number.[11]

Through the Darkazanli investigation and the Zammar surveillance, at least by the end of 1998, German secret services knew the names of all the men in the Amir-Omar group. They saw nothing suspicious, or at least nothing illegal.

"We only knew them as radical Muslims. This is not a crime," one investigator said.[12] "They might have had contact with followers of Osama Bin Laden. This is also not a crime."

To recruit people for jihad was not uncommon, unprecedented, or illegal. For more than a decade, thousands of men throughout Western

Europe went to Afghanistan, Bosnia, or Chechnya to fight or, more usually, to the Afghan training camps as a sort of baptism into the broader goals of radical Islam. It became, within that world, a praiseworthy, almost unexceptional thing to do.

Security agencies in Europe and the United States usually viewed men such as Zammar as individuals, not as part of formal networks. There were, however, some formal fundamentalist networks actively at work recruiting young Arabs. In some instances, these networks overlapped with—and took advantage of—a missionary sect of Muslims called the Tabligh. The Tabligh, often likened to a Muslim version of the Jehovah's Witnesses, proselytizes throughout the world. It professes to be peaceful, but intelligence services throughout the Mideast said the group's purposes were sometimes hijacked by organizations such as Al Qaeda to recruit mujahideen. Sometimes, people purportedly speaking in the name of the Tabligh—would recruit men to come on religious missions, then deliver them to training camps for different fighting groups.[13]

Unknown to German investigators, Zammar was a Tabligh.[14] Some years before he had traveled to Pakistan at the group's invitation and joined. The organization controlled one of the three Arab mosques in Hamburg, Al Nur Mosque, and Zammar often used its premises to meet potential recruits. He did the same at Al Quds and although he had no formal role at any of the mosques, he was a constant presence, urging young men to dedicate themselves to the cause.

He was a frequent guest at the apartment house in Harburg, too, pulling up on the treeless street fronting the flat and hauling cartons of books and recruiting materials upstairs.[15] He almost always wore a traditional Arab tunic and a checked scarf. He also made the rounds of Amir's study groups and organized social outings for the younger men.

In Hamburg, as in other German cities, some residents rent parcels of land on the urban outskirts or along freeways and rail lines and use the plots as vegetable gardens. They frequently build small huts on the parcels, and some use the huts as working-class rural retreats. Zammar's family had a hut like this near Stadtpark, the city's largest green space, and he sometimes invited members of the study groups out for a barbecue—a cheap holiday.[16] His approach seldom varied: He urged the young men to commit themselves to the cause, to do as he had done, to go and fight.

Zammar's bluster proved an effective disguise. The Germans discounted the idea that anyone so public in his displays of affection for radical Islam would actually be doing what he seemed to be doing—recruiting for radical Islam. But he was—and the men in Harburg were increasingly receptive to the call.

When the security services, in the course of their surveillance of Zammar and Darkazanli, became aware of the Harburg group, they looked at its members but quickly concluded they were fundamentalist Muslims with no political connections. They entered the names of Mounir el-Motassadeq and Said Bahaji on watch lists of radicals whose movements ought to be monitored.

German law was enormously respectful of the individual's rights of privacy. The shadow of national socialism and the Nazis' routine abrogation of rights informed a set of protections against state power that make American civil liberties pale by comparison. News organizations, for example, are restrained in many instances from publishing or broadcasting the full names of criminal suspects, or even convicts. The fact of a criminal conviction, even a heinous one, is not a matter of public knowledge. This attitude runs deep in the legal code, and bringing criminal charges against individuals because they are suspected of belonging to terror organizations wasn't possible. The result was that no matter what suspicions German investigators might have had, they were reluctant or unable to proceed against people classified as political extremists.

So it was that men like Zammar and Darkazanli, despite evidence of connections to foreign terrorists, remained at large and, at least in Zammar's case, continued to engage in the activities that caused suspicion to fall on them in the first place. This drove some American counterterrorism specialists crazy. The Germans had initially resisted executing the warrant that led to the arrest of Bin Laden's finance man, Salim. And even after Darkazanli had been implicated by information gathered from Salim, the Germans felt they could do nothing against him. There was no evidence of criminal activity, they said, at least as defined by German law. Legally, there was very little the Germans could do.

This was the situation that confronted Thomas Volz when he was posted to the American consulate in Hamburg in 1998, ostensibly as a commercial attaché. Volz, in fact, was a veteran CIA field agent. The agency had been aware of Darkazanli since 1993 and was convinced he was an Al Qaeda member,[17] but it had been unable to build a case against him. The information gathered after the embassy bombings reinforced the Americans' belief, and they repeatedly urged the Germans to do something about him. Volz decided that if Darkazanli couldn't be arrested, the best thing to do would be to "turn" him, make him into an informer. He could be bought off, Volz told his German counterparts.[18] The Germans laughed. Money means nothing to true believers like Darkazanli, they said, and declined to pursue Volz's proposal. Volz himself could do nothing without violating the informal agreement between the two countries' secret services, so was reduced to shopping his proposal to different German agencies, all of which thought it was a typically brazen—which is to say, stupid—American idea.

After continued insistence by Volz and mainly as a way to shut him up, the Germans eventually approached Darkazanli, who, the Germans said, brushed them off.[19] Volz continued to agitate, and at one point was threatened with expulsion from the country if he didn't calm down. The net result was Darkazanli was neither turned, arrested, nor, apparently, constrained in any way.

TUHH

After the move to Marienstrasse, Amir, who had done nothing toward his graduate degree in a year, suddenly called his TUHH faculty adviser, Dittmar Machule, to plan completion of his thesis. Machule asked him: "Where have you been, Mohamed? . . . There is trouble? Problems in the family?"[20]

"Yes, in the family, at home," Amir told him. "Please understand, I don't want to talk about this."[21]

And that was that. It was as if he had never left. Amir began vigorous work on his Aleppo thesis. He returned to the university in other ways, too. He, Bahaji, and a third student, a Pakistani air force officer, petitioned the university for space for a prayer room. In his appeal, Amir said not having a place to pray was an undue hardship. They were granted the request, and it gave Amir yet another place where he could teach others about Islam.

Amir resumed regular meetings with Machule to discuss his ideas and progress, and six months later, in the spring of 1999, he turned in a 152-page manuscript. Machule opened it to find an inscription and dedication to Allah on the first page. The rest of the work held few surprises. It was a solid, thorough examination of Aleppo's history and current redevelopment and a proposal to better integrate the city's past with its future. Machule judged it to be of high quality intellectually but uneven in its writing. He asked another professor, Chrilla Wendt, to work with Amir to polish the thesis before it was formally submitted. Wendt and Amir worked together at regular meetings, side by side at a desk every day for six weeks. Wendt knew of Amir's discomfort around women but said the work went smoothly until, suddenly, at the end Amir told her that he could no longer bear to be in such close proximity. By then, the rewriting was nearly complete, and in August, Amir formally submitted the thesis. He defended it before a departmental review committee later in the month and received high marks, and congratulations. The degree came later in the mail.

The thesis was a routine piece of urban analysis, albeit from an Islamic perspective. One of the concerns Amir addressed in it was how to plan contemporary cities to accommodate women being at home all the time, not working. The most interesting thing about the thesis overall was Amir's decision to finish it at all. Why, after having virtually abandoned his academic life, did he suddenly return to it with such dedication? No one ever knew. Amir never talked to any of his instructors about his intentions.

Machule remembers Amir coming by his office one final time after he had been awarded his degree. Machule was busy talking with another student and Amir, being Amir, didn't barge in. He didn't even knock. He just stood at the open door, hoping to catch Machule's eye. Machule saw him and gestured for him to wait. Amir did; he stood at the door, gaze averted, silent, for another ten minutes. Then he turned and walked away, and Machule never saw him again.

Whatever he intended to do next, it seemed clear it would occur somewhere else. He e-mailed his old research and study companion, Volker Hauth, to tell him the news. The note showed quite a different Amir, light and playful, almost cute. Maybe it was the simple relief of having finished those many years of study.

Hello Volker,

Heh you unfaithful tomato. Do you still know me? I am Mohamed El-Amir. I finished my studies (finally). I will leave Germany soon. Here is stuff, that belongs to you and something else, I want to send you. I wish you all the best in your life. I am sure we will see [each other] again.

So, till then.

Mohamed el-Amir, August 31 1999.[22]

In addition to Amir's academic work, a lot of other loose ends were being tied up. Several members of the Harburg group executed powers of attorney, assigning control of their bank accounts and other finances to acquaintances. Four members of the group married in a period of six months: In the spring, Ziad Jarrah married Aysel Sengün at a ceremony at Al Nur Mosque in Hamburg; in the summer, Marwan al-Shehhi returned home to the Emirates to marry a woman from his hometown chosen by his family; in the fall, Mounir el-Motassadeq married a Russian émigrée and Said Bahaji married a local Turkish woman.[23]

Jarrah's wedding must have been a desultory affair. It was done over Aysel's objections not at Al Quds, where Jarrah worshipped, but at the Tabligh mosque. It was almost as if he were trying to hide the event. Neither family was informed, before or after. It was clear the wedding was intended to do little more than placate Jarrah's friends, or perhaps his conscience. Jarrah won this little war, but it's hard to say what the spoils were. They lived apart, separated by far more than the geography of northern Germany. Their visits to one another grew further and further apart. Their contact between visits was sporadic. The couple never registered the marriage with the state, and Aysel said she didn't consider it genuine. It was real enough, however, that she insisted the two of them sign a contract beforehand that specified she could continue her medical studies. Jarrah later tried to renege on this and asked her to quit dental school, but she appealed to the imam who performed their wedding ceremony and he backed Aysel's position.[24] Jarrah complied, but reluctantly; he never stopped complaining about her independence.

They broke up again within weeks of the wedding, and then, as usual, quickly made up. Jarrah, for a man given to dark moods, had a very lighthearted, loving touch at the keyboard. He frequently signed his notes to Aysel with a long drawn-out goodbyeeeeeeeeeeeeeeeeeeeeee, followed by multiple exclamation points. In May, after they had reunited, Aysel wrote to him: "It's me again. How is my darling? All I can say for me is I miss you very very much. Meow. I want to cuddle. I love you."[25]

In June she wrote, "I just wanted to write you that I'll always love you. I'll contact you as fast as I can. Your Aysel." Another time that spring she signed a note, "I love you, your yearning wife, Aysel Jarrah."

By summer, they had broken up again.

"I thought it is forever, and he probably, too, but we got back together on the telephone after two weeks," Aysel said.

After Jarrah had moved to Hamburg, he had grown increasingly distant, not just from Aysel but from his family in Lebanon as well. It was a close family; he was the only son and they grew concerned about him. Jarrah fell behind in his studies. Friends told the family he spent most of his time in the mosques and had begun making regular donations to them as well. They sent emissaries to talk to him and threatened to cut off the generous monthly allowance his father sent. When that didn't work, according to investigators, his father once feigned a heart attack in hopes that Jarrah would come home. He didn't.

Jarrah was never a central member of the Amir-Omar group. He went to the University of Applied Science, which was in the central part of Hamburg, not convenient to Harburg, a southern suburb, where the Marienstrasse apartment and the TUHH were located. Jarrah was seen occasionally in Harburg, but not regularly. He was close to Omar but seemed to barely know Amir at all. The other men usually lived in various combinations with or near one another, but Jarrah, when he first moved to Hamburg lived for several months with Bashir Musleh, his friend from Greifswald, and from then on rented a room on his own.

Jarrah saw less and less of Aysel. She visited Hamburg a few times but complained she felt unwelcome; on at least one occasion, Ziad abandoned her in his rented room to spend time with his friends, whom she never even met. She was angry, but Jarrah said that where he went, women could not go.

"He spoke about religion in general and tried to convert me step by step," she said. "He enlightened me about the problems Muslims have in the Middle East. He also spoke about the intifada. I wouldn't have known what the intifada meant at that time because I don't have a political background. When I asked, Ziad explained it was the freedom struggle of the Palestinians against Israel."[26]

Aysel later told investigators that she first heard Jarrah talk about jihad just after he moved to Hamburg in late 1997.

"I didn't know what it means," she said. "I asked Arab friends about the meaning. Somebody explained to me that the word 'jihad' in the softer form means to write books, tell people about Islam. But Ziad's own jihad was more aggressive, the fighting kind, giving oneself up for the religion. . . . He meant your personal interests were not as important as the religion. His opinion about the jihad, the holy war, I was afraid because of that. That was the reason I spoke to friends about it, to learn more about the jihad. And I also felt at that time that I wasn't the center of Ziad's life anymore. It was faith and religion. He started to visit me less. Because of me being afraid I did not mention the issue of jihad anymore. I couldn't and I didn't want to understand and I couldn't hear of it anymore."[27]

The couple broke up and reconciled over and over. During one of the reconciliations, Aysel became pregnant. She told Jarrah about the pregnancy, then aborted it, she told investigators, because of the uncertainty of their relationship. She later apologized to Jarrah by e-mail: "I had to think about our baby today. I am sorry about everything I did to you."[28]

It was clear to Aysel, at least, that Jarrah's life had taken a very serious turn. She told a friend: "I don't want to be left behind with the children, because my husband moved into a fanatic war." Aysel at one point contemplated moving back with her parents in Stuttgart. Instead, she transferred to the University of Bochum, in coal country near Düsseldorf. Theoretically, it was more convenient to Hamburg—it was a straight shot on the train—but Jarrah's visits remained irregular. Aysel, as even she admitted, was an almost excessively social woman. The isolation was killing her. She would grow frantic from loneliness and her frequent inability to track Jarrah down. Once, in desperation, Aysel wrote him:

> Again you haven't been reachable. I left a message for you to call me back. Since you haven't done so, I assume you haven't been home at all. I couldn't sleep last night and I thought for a long, long time. What is love for you? . . . I want to tell you what love is for me: To take the other as he is, to share everything with him you have (mentally and physically, materially, in all areas of life) to do something for the other you wouldn't do for yourself, to be there for the other (especially in bad times). . . . I just want to ask you one thing: Be honest to me, don't just say it, if you don't mean it with all you believe and if you think I would change my mind about jihad. . . . Think carefully about it, if you can't give me that promise, it is better to forget about our marriage even though it would hurt a lot. . . .
>   In love. Aysel.[29]

When Jarrah was out of touch, Aysel would try to track him down, calling all the numbers she had for him or any of the friends he had allowed her to meet in Hamburg. When that proved unproductive, she would call anybody else she thought might know him. Once, she grew so desperate to find him she combed through her old telephone bills for calls Jarrah made from her flat when he visited. She called every number she didn't know, demanding that someone tell her where he was. It was fruitless. For the first two years in Hamburg, Jarrah maintained, at least nominally, his commitment to Aysel and he dutifully kept up his studies in aeronautical engineering. Gradually, however, he began to fall away, further from her and deeper into his new world, the one where she wasn't allowed to follow.

He told her once that he was ashamed of her, of her Western ways, smoking, drinking, acting like anything but a dutiful Muslim wife. Aysel's old roommates said the criticism sometimes grew violent. Once, she told them, he hit her. He threatened to do worse to one of her German roommates, with whom he had become friends. "Today I am sitting here with you," he said, "and tomorrow I will kill you."[30]

KAFR EL-SHEIK

Said Bahaji had no such problem with his prospective wife. She was the devout daughter of an imam at one of the Turkish mosques in town, as comfortable with Bahaji's fundamentalism as he was. They married in October at Al Quds. Brother Haydar Zammar was there; Omar gave an overtly political lecture, apologizing for bringing politics into the ceremony, then forging ahead and doing so; and Marwan al-Shehhi led the singing of jihad songs. One striking aspect of traditional Muslim weddings is that the bride is nowhere to be found. Bahaji's wedding, in that respect, mirrored the life these men were living. They were alone in the mosque in one another's company without intrusion from the world.

"We are now in school, like in Arabic lessons," Omar said. "In the end we will have a test. In this test some will pass and others won't."[31]

It sounded like a warning, but as usual with Omar, it was delivered without anger and with apparent good cheer. He even paused a couple of times to smile. He went on to denounce the Jewish occupation of Jerusalem and to remind those present of their responsibility to end that occupation. "The problem of Jerusalem is the problem of the [Muslim] nation. That's also the problem of every Muslim everywhere. Whenever possible and during each jihad a Muslim has to remind his comrades about that. He has to remind him of the problem of the nation, the beloved nation. To talk about that does not harm this wedding, quite the contrary. Every Muslim has the aim to free the Islamic soil from the tyrants and oppressors."

Omar quoted a poet to the effect that Jerusalem would one day be swept "by a wave of fire and blood" before relinquishing the floor to Shehhi and his friends who sang jihad songs. One went:

I came to this life
Which is only a short elusive pleasure
A journey through, a battle
I became fire and light, a melody and fragrance.
Until I lived a generation
Which I spend watching you through the light
My eyes are full of light
Mine are the virgins in paradise
I sing like an angel
That you are the light of my eyes
These gardens smell sweet
And their smell is my wound
He is the spirit
I have been visited by the prophets
And my brothers are martyrs

Preaching holy war and martyrdom at the wedding did not seem to upset anyone there. In fact, at one point the audience, unprompted, broke out into a call-and-response jihad chant. The men ate dates and carefully placed the pits in little candy dishes passed through the room and shouted for war. They hugged one another and placed the ritual three kisses on one another's cheeks at the end of it. Young boys played in the back of the room and off on the far side of the room there was hung a cream-colored curtain. Behind it were the women, although they never emerged.

Mohamed el-Amir missed Bahaji's wedding. He had gone home to Egypt, instead, to visit his parents. His father greeted him as a conquering hero in Cairo and began the task of trying to find him a bride.[32]

"I told him we should look for a wife for him," Amir's father said. The elder Amir was a skilled lawyer. He always had his case prepared. This time, he had a potential bride already lined up: "We went to visit a family, and Mohamed met the daughter and they liked each other. The woman's parents also liked Mohamed, but their only condition was that their daughter not leave Cairo. So Mohamed got engaged to her and then went back to finish his Ph.D."

At least that was what Amir had told his father, that he was going to go from Germany to the United States to pursue a doctorate. He told friends the same thing. Amir's parents were, by this time, estranged from one another. There had been a dispute over arrangements for their eldest daughter's marriage, according to Amir's aunt, Hamida Fateh. Amir's father didn't approve of the groom, a heart surgeon, who had been selected by the mother's family. Bouthayna's health—she had diabetes—had been in decline, and Amir's visit was a time of great solace and joy for her. She took him north to the Delta, to Kafr el-Sheik, to show him off to her relatives. "It made her very, very happy," Fateh said.

Amir looked fit and handsome, Fateh said, ready for anything, but he told his mother he was unsure about continuing his education. What he really wanted to do was stay; he said he was tired. He wanted to remain in Cairo and take care of her. He asked if she would encourage his father to allow it. Bouthayna would not. She insisted that his father was right, that he must continue with school.[33] You need to get a doctorate, she told him; go to America.

HAMBURG

The Arab community in Hamburg was small enough that, especially among the fundamentalists, everybody knew everybody, but relationships among them were constantly evolving. The young Islamists in Hamburg never formed a large static group. They formed a series of overlapping smaller groups who moved in and out of one another's orbits. The same could be said of their beliefs.

The young men pored over religious texts. They parsed videotapes of sermons by different preachers. One video they had was a sermon by the London-based Palestinian imam known as Abu Qatada,[34] one of the principal theorists of radical Islam. He was heavily networked into local movements in several countries, and his London mosque was a magnet for activists from all of them. In the Hamburg video, he implored believers to throw off the yoke of the infidels who ran the world, kill their children, capture their women, and destroy their fields and homes. God has already made known his verdict on them, he said, and they deserve to die.

This was the kind of idea the Hamburg men had debated for years among themselves. There was a notion current among fundamentalists, originating among Egyptian theorists, which held that Muslims had a "neglected duty," or *al-faridah al-ghaibah*, to make right the world, and to do whatever that required.[35] Many among the Hamburg men were shocked that a Muslim holy man would advocate the behavior Abu Qatada called for. Many of them knew, also, that holding ideas such as this would land them in jail, or worse, in their home countries. The expression of such radical ideas, in fact, was far more common—and much more public—outside the Islamic world than within it. Arab rulers kept constant watch for signs of Islamist tumult; they danced around it publicly, one month tolerating it, the next crushing even the merest hint of it. Governments in the West, apparently thinking they had little to fear from such sentiments, often ignored it entirely. Or, if they made note of it at all, they tended not to act to suppress it. This was what happened in Hamburg. The increasing radicalism of the Islamist scene was well known to authorities who did nothing to interfere with it. Over time, the men who cautioned against the sort of ideas endorsed by Abu Qatada, or the many others like him, drifted out of the Hamburg circle. Some finished their studies and left town. Others moved to different universities in Germany. Some simply left for fear of where their friends' ideas seemed to be taking them. The effect was a distillation of both the men and the ideas. Those who stayed tended to be the most radical and they tended to reinforce one another's most radical ideas, driving them toward the furthest reaches of Islam.

The central idea that began to dominate their discussions was the notion of jihad. "Paradise," Amir and Omar reminded their friends, "is overshadowed with swords."[36]

Within Islam there are conventionally two distinct notions of jihad. The first, and historically predominant, is the individual's daily struggle for his own soul. This sense of jihad is not much different from core beliefs held in many other religions; it is what has made religion a normative force in most societies and led it often to be praised for its socializing effects even by nonbelievers. The prophet Mohammed said: "The best jihad is the one who strives against his own self for Allah, the Mighty and Majestic."

The second form of jihad is to fight physically against the enemies of Islam. The Qur'an states: "Fight in the cause of Allah those who fight you, but do not transgress limits; for Allah loves not transgressors." This traditionally was interpreted to mean Muslims must defend themselves and their lands, but should never be aggressors. Radical Islamic thinkers in the twentieth century advocated a greatly expanded notion of what constituted defense. In essence, they said, all non-Muslims are the avowed enemies of Islam and would destroy it if they had the opportunity. Muslims who were insufficiently devout, including those who ruled much of the Arab world, were also enemies of Islam. These thinkers, and those who followed them, reinvented jihad as an offensive weapon. This newly ascendant idea of jihad was the motivating cause of radical Islam. The individual Muslim had an *obligation* to fight on behalf of his beliefs, against nonbelievers and corrupters of belief.

This was the great awakening for modern Muslims. Osama Bin Laden, a veteran of that fight, codified the obligation after the Afghan war: Muslims were on the defensive against the aggression of the Zionist-Crusader alliance; jihad was the duty of every Muslim. To kill Americans and their allies, both civil and military, was an individual duty of every Muslim who was able, in any country where this was possible, until Al Aqsa Mosque in Jerusalem and Al Haram Mosque in Mecca were freed from their grip and until their armies, shattered and broken-winged, departed from all the lands of Islam, incapable of threatening any Muslim.[37]

Zammar, also a veteran of the Afghan war, constantly spoke in favor of jihad; and those within the Omar-Amir group, especially after Marwan al-Shehhi came along, began to embrace the idea. Omar, although he had always spoken favorably of jihad in theory, had at times expressed some skepticism about personal participation; he would hem and haw when Zammar urged the men to move beyond contributions of mere money or moral support to their brothers in Bosnia or Chechnya. But that skepticism faded over time and he began to romanticize the idea of fighting; he imagined himself in a mud trench in battle.[38] The original followers of the prophet Mohammed were Yemeni, and there is some sense among

Yemenis now that they are the true believers, the chosen people waiting to be called back to defend the throne. Omar typified this intuitive, unthought faith. The quality of Shehhi's belief was similar—received and literal rather than reasoned. He spoke longingly about paradise, sitting beneath a lone shade tree on the banks of a wide river of honey. Amir had a much more contemporary relationship with his religion. His crabbed personality not withstanding, he was a modernist, a rationalist—an engineer—to his bones. He thought you could analyze anything and come to the right answer. Once he had arrived at an answer, he was unshakable, but he had to get there on his own. In his discussion and prayer groups around town and in the little flat in Harburg, Amir was careful not to press others to agree with radical formulations. "No force in faith" was one of his guiding principles.[39] In the end, there was the Qur'an and the individual's interpretation of it. The individual, not the group, must answer to God for his actions or lack thereof. He even took care to arrange a private meeting with one of the younger men in his study group and warned him to stay away from extremists who might seek to shape his thinking.[40] Follow the Qur'an strictly, he told him, but live a careful life.

Normally, this defense of the individual's right to choose gave the group's discussions a pious, polite, and restrained tenor. But as other moderating voices drifted or were chased away, Amir, Omar, and Shehhi began to grow more and more agitated whenever the conversation turned to jihad, the Jews, and America. It was the only time they got truly excited. Sometimes, out of nowhere, one of the men would call out, "Our Way!" The others would answer with a shout: "Jihad!" Then they'd repeat it in a chant: "Our Way! Jihad! Our Way! Jihad!" Because they frequently stayed up talking into the early hours of morning, neighbors complained about the noise every place they lived. The young men began meeting regularly with Mohammed Fazazi, the ferocious Moroccan imam who urged relentless jihad in its most violent forms. Fazazi preached for months at a time at Al Quds and took a particular interest in the group, whose members seemed more serious, more focused than many other young fundamentalists.[41]

The group endorsed the words of the Prophet in which he likened the community of Muslims to a body, and if Muslims anywhere were hurt or attacked, it was as if the body had a fever. Not to fight to cure the fever was a sin.

The jihad, they thought, would eventually purify the earth, would put an end to the *fitna*, or confusion, and sacrilege that was everywhere apparent. Muslims had been beset by this confusion and it was the responsibility of those who were clear of thought to free those who were not. Amir, in particular, hated the lack of order he saw in the West. As far back as his first year in the university, he had reacted to a typically boisterous student gathering—the pre-film conversation in a movie theater—by sinking down in his seat and muttering, "Chaos, chaos." Amir told his own students the answers offered in the Qur'an were clear and simple and would, if you read and carefully considered them, clear the way. That was where he saw the necessity of jihad—to get rid of the Jews and Americans who intentionally induced misunderstanding. The ways of the West, he said, were the means by which Satan tempted and spread confusion among the believers.[42] It was every Muslim's duty to fight this.

The path before them had become clear: they must, as the Prophet had said, take up the sword.

AFGHANISTAN

In early October, Ziad Jarrah, in a note to himself, wrote: "The morning will come. The victors will come, will come. We swear to beat you. The earth will shake beneath your feet."[43] And a week later, just days before Aysel was due to arrive on the train up from Bochum for the weekend, he wrote: "I came to you with men who love the death just as you love life. . . . The mujahideen give their money for the weapons, food, and journeys to win and to die for Allah's cause, but the unhappy ones will be killed. Oh, the smell of paradise is rising."

Shortly after, Aysel visited him in Hamburg. The visit did not go as Aysel had planned. They seldom did, but this was different. The wedding had resolved nothing. Their disagreements were as sharp and many as ever. Jarrah always could be distant. Now he was absent. "That scared me," she said. He had begun talking a lot about Chechnya, and he seemed weighed down by some great burden he was carrying. Aysel suspected something momentous was about to happen, but she couldn't tell what. He had quit going to classes. He told Aysel he was going home to Lebanon to consider where he was in his life and where he wanted to go.

Together, they cleaned out his small room, packed his things, and divided those that he would take home to Lebanon, those he would leave with Aysel. He stored his college papers and textbooks at Bashir Musleh's student apartment. When he took Aysel to catch her train back to Bochum, she was filled with dread that she would never see him again. Aysel knew that wherever he was going, it almost certainly wasn't home to Mom and Dad.

She was right. That month Jarrah, Amir, Omar, and Shehhi began making arrangements to travel from Hamburg through Turkey to Pakistan—the Zammar route. Shehhi withdrew $7,000 from his bank account to pay for the air tickets.[44] Shehhi granted power of attorney over his accounts to Motassadeq. He also asked Motassadeq to cancel his apartment lease. He told him he was going back to the Emirates to be with his wife and wouldn't be returning.

They traveled separately, so as not to attract attention, either by their sudden absence or en route. Shehhi left first, then Amir and Jarrah. After Amir left, Shahid Nickels asked Omar where he was, why he wasn't teaching his classes.[45] Omar said Amir had gone to America to pursue his doctorate. Nickels asked again a week or so later, and this time Omar smiled and said, No, the U.S. didn't work out. He's in Malaysia. Then Omar left, too.

BOOK TWO

The Engineer

CHAPTER 1

The Rebirth of Jihad

PESHAWAR

IN JULY 1973, IN KABUL, AFGHANISTAN, Daoud Khan led a bloodless coup that overthrew his cousin, the King of Afghanistan, who was vacationing in Europe at the time. Daoud installed a new government controlled by his Afghan Communist Party. Although it seemed of little import outside Afghanistan at the time, the coup precipitated three decades of almost ceaseless warfare that eventually drew nearly the entire planet into its orbit. Daoud was himself deposed, and killed, in a second coup five years later. After that coup produced yet another the next year, this one accompanied and followed by waves of violence and bloodletting, the place seemed to be spinning out of control. More to the point, it seemed in danger of falling out of the grasp of its near neighbor to the north, the Union of Soviet Socialist Republics. Airborne commandos from the USSR landed in Kabul in December 1979, the heralds of an armed force that eventually numbered more than 100,000. The government was overthrown again and yet another new one installed. The Soviets spent the next decade fighting what became, in effect, the last stand of the empire.

Their outright invasion into Afghanistan was widely condemned throughout the community of nations, but nowhere with the vehemence expressed in the Muslim world, where it was regarded as a direct assault on Islam, one that required a direct response. The herald of that response was a Palestinian academician named Abdullah Azzam.

Azzam, born in 1941 in Jenin on the West Bank, was a product of the Palestinian diaspora, steeped in Islamic learning and politics. He answered his first call to battle in the 1967 Six-Day War, after which he joined the Palestinian resistance, and later left because it was, he said, merely a political cause insufficiently rooted in Islam. He joined the Muslim Brotherhood and later helped found Hamas as an Islamic alternative to the Palestinian Liberation Organization. He resumed an academic career that earned him degrees in Jordan and Damascus and eventually a Ph.D. from Cairo's Al Azhar, the high temple of Islamic learning.

When the USSR invaded Afghanistan in 1979, Azzam was among the first of the non-Afghan Muslim sympathizers to join the cause against the Soviets. He came at once to Pakistan, initially to the capital, Islamabad. When that proved too distant from the battlefield, he moved his base to Peshawar, capital of Pakistan's untamed Northwest Frontier Province, a land governed chiefly by fear and high-caliber ammunition. There, Azzam found his cause: Afghanistan would be the incubator for a new, muscular Islam, a religion of warriors like that of the Prophet's time. The core texts of the standard modern Islamist creed had already been written by Egyptians Hasan al-Banna and Sayyid Qutb, one the founder and the other the great popularizer of the Muslim Brotherhood. Remaking old teachings for the modern world, they sought the imposition of Islamic law in their native Egypt and any other country that pretended to be Muslim. To achieve their ends, they insisted on violent revolt in the name of God, against the unfaithful and against apostate Muslim regimes. Their insurgency was greeted with fierce, deadly, and largely effective repression within Egypt. Because of the response, it remained mainly theoretical and contained within the intellectual elites until Afghanistan. There, Azzam saw a chance to make real the abstract promises of Qutb and Banna, to fight an unequivocal battle for Islam and, in doing so, to recover their virtue. "One of the most important lost obligations is the forgotten obligation of fighting," he said. "Because it is absent from the present condition of the Muslims, they have become as rubbish of the flood waters."[1]

Jihad, Azzam wrote, was the way of everlasting glory, and the only way to get there was behind the barrel of a gun. "Jihad and the rifle alone: no negotiations, no conferences, and no dialogues," he said.[2] Azzam, more than any man, popularized the modern—and, within its world, triumphant—notion of the contemporary Muslim's duty to wage holy war. The goal was no less than to resurrect the reign of Islam on Earth, and its imperative was universal. It was a duty, Azzam said, that commanded all Muslims to its banner. He quoted the Qur'an:

> Then, it is obligatory upon the whole of creation to march out for jihad. If they fail to respond, they are in sin. . . . The light, the heavy, the riding, the walking, the slave, and the free man shall all go out. Whoever has a father, without his permission and whoever has not a father, until Allah's religion prevails, defends the territory and the property, humiliates the enemy and rescues the prisoners. On this there is no disagreement. What does he do if the rest stay behind? He finds a prisoner and pays his ransom. He attacks by himself if he is able, and if not he prepares a warrior.[3]

Azzam's new home, Peshawar, sat at the head of an upland basin ringed by hills that rose into mountains north and west en route to the Khyber Pass and Afghanistan beyond. It had always been more a city of Pashtuns—the dominant ethnic group in Afghanistan—than Pakistan, and had in fact been the capital of Afghanistan at one time. The Pashtun domain is a region of harsh landscapes and harsher lives, with not much available arable land or possibility of livelihood. There is an unsettled, Wild West quality to the region, and the Pashtuns were often seen by outsiders as very wily characters, as hard as their lands and not to be much trusted. Trust, the Pashtuns said, was not the issue. They lived by a simple code: Covet a man's women, gold, or land, and you would almost certainly have to deal with the man himself. One of you was apt not to survive such an encounter.

Over the centuries, Peshawar had evolved into a more sophisticated place than its hinterland, owing mainly to its location on the ancient Silk Road linking Europe and Asia; it became an international crossroads for traders, warriors, and backwoods statesmen. Like many entrepôts, its history grew rich with tales of compromise, dispute, and war. In the 1980s, Peshawar became the capital of the Afghan resistance. With or without the Soviet Union, Afghanistan's internal politics were complex, layered with historical, tribal, ethnic, ideological, and personal antipathies that hadn't been resolved for most of a millennium.[4] The immediate effect of this history with regard to the Soviets was an internal resistance to the invasion that was every bit as layered and complicated, and frequently as antagonistic to itself as to the Soviets.

Even before the Soviet invasion, a civil war had been underway. The Soviet purpose was to change one puppet government for another, one they hoped would be more palatable to the Afghans yet still compliant to Moscow. Domestic opponents of the government—in particular, those who thought it lacked Islamic purity—had already begun establishing exile organizations based in Peshawar.[5] After the invasion, the number of those organizations—not to mention exiles—mushroomed. In the beginning, most of these groups were under-funded, or not funded at all, and survived mainly on what they could steal or conscript. But funding, more bountiful than anyone could possibly have imagined, soon arrived. By the middle of the 1980s, more than $1 billion a year was pouring into the resistance. Sorting out the varied Afghan opposition was an almost impossible task, but sorting out who would get the money was imperative. A hierarchy was constructed at the insistence of the chief sponsors, mainly the United States, Saudi Arabia, and, especially, Pakistan. The Pakistanis, who regarded themselves as the general staff of the war effort,[6] designated six Afghan political parties to receive money and later added a seventh. All were fundamentalist religious parties and all eventually fielded separate armies.

As the geographic headquarters of the resistance, Peshawar naturally became the destination for those from beyond Afghanistan who wished to join it. Largely through efforts initiated by Azzam, thousands of Muslims came to join the Afghans in holy war against the heathen Communists. Azzam internationalized the jihad. He worked tirelessly—and globally—as a recruiter. He traveled to Europe and throughout the Middle East and made several trips to the United States, raising money and, at least theoretically, an army to fight the jihad. Funded by donations mainly from

Saudi Arabia and other Gulf states but with contributions from around the globe, he opened recruiting offices in thirty countries, including the United States. His writings were not intellectual treatises, but pure agit-prop; Tom Paine, not John Locke. His activities in the United States were endorsed by the American government, which chose to ignore its content while actively promoting its results.

The United States was one of the main allies of the Afghan resistance. President Jimmy Carter had earlier confronted the Soviet Union in his human rights campaign and his national security adviser, Zbigniew Brzezinski, was eager to confront them with surrogates on the battlefield as well.[7] Acting on Brzezinski's recommendation, Carter had approved American financial and military assistance to the Afghan resistance even before the Soviet invasion. Brzezinski said the express intent of the secret assistance was to encourage Soviet intervention, which of course it did. The Afghan resistance, Pakistan, Azzam, and the Americans were joined in their unlikely alliance by Saudi Arabia, which provided hard cash and volunteers.

"Muslim governments made it a religious issue, a way to reinvigorate Islam, to counter the Western cultural invasion, to counter Iran and the Shiite threat to Sunnis," said Waheed Hamza Hashim, a Saudi political scientist. "So here comes the Soviet invasion in the midst of this. The Arab world sees itself under siege. The government's main legitimacy derives from religious authority. They see this as an opportunity to strengthen their religious credentials, thereby strengthening their political authority. To defend against Iran, mobilize religiously. To defend against the Soviets, mobilize religiously. 'We're going to jihad. Want to come along?' When the mobilization took place, it was not institutionalized; it was localized. Everyone was empowered. Everyone took initiative.

"A businessman wanted to give *zakat* [charity] to Afghanistan. He goes and gives it himself. Gets on the plane and goes to Afghanistan. The individual has the power, the responsibility. Everything was hectic. There was no way to organize it. There was no way to control it. The beast just grew and grew."[8]

Each of the allies—the Afghan mujahideen, United States, Pakistan, and Saudi Arabia—opposed the Soviets for its own reasons: the Afghans, most obviously, wanted to reclaim control of their country; the

Americans, as they had been for thirty years, were principally focused on *who* they were fighting—the Soviets—not where; their concern was not Afghanistan but superpower realpolitik; the Saudis were protecting their domestic base of power, which rested primarily on the government's alliance with religious authorities; and the Pakistanis wanted to control Afghanistan as both a buffer state to the Soviet threat from the north and a redoubt to attack India from the south. The Saudi government sent hundreds of missionaries and billions of dollars. The United States funneled arms through Pakistan and matched the Saudis million for million, most of it moving eventually through Peshawar.

Gun violence had been a way of life and death in the region long before the Soviet war. There is a Pashtun saying: A man's jewelry is his gun. But there had never been anything on this scale. Pakistan's powerful Inter-Services Intelligence service (the ISI) was installed as executor of American and Saudi interests. The ISI was at the time—and remains—the most powerful organization in Pakistan. To a significant degree, the ISI has made and unmade governments in Islamabad since the country's separation from India. Its leaders had their own motives in Afghanistan, which were religious as well as political. The ISI was among the most highly Islamicized organizations in Pakistan and many of its leaders wanted an Islamic client state established in Afghanistan. Their pursuit of their objectives—at times in contradiction with those of the United States and Saudis—created confusion and difficulty in the administration of the war from the very beginning. In choosing the ISI to serve as the general staff of the war, U.S. officials recognized this as a danger but knew also that they had little choice. The service created a new logistics operation, a sort of third world UPS, just to distribute the flood of armaments. And there was nothing third world about the weaponry. Convoys of 10-ton trucks filled with automatic rifles, machine guns, grenade launchers, and, by the war's end, most cherished of all, Stinger antiaircraft missiles were sent out daily on the cross-country trip from the Arabian Sea docks in Karachi to Peshawar and the Afghan interior. At least that was the intention. Much of the armament was "taxed," or simply stolen, before it ever reached the border.[9]

The overwhelming number of those who answered Azzam's recruitment call to, as he put it, join the caravan were Arabs. It was, said General Hamid Gul, who headed the ISI at the time, "the first international brigade of the modern time. The Communists had their international brigades, the West has NATO, why should we Muslims not unite to one common front?"[10]

Pakistan granted visas carte blanche and gave directions to the front to anyone from anywhere who wanted to join the fight. Saudi Arabia's national airline offered special half-off jihad fares from Riyadh and Jeddah. Arab governments sent emissaries and opened offices for dozens of state-sponsored charities to provide humanitarian relief and, no less important, to assist the fighters. For a decade, parts of Peshawar were transformed into a sort of Little Mecca.

"It was a bustling Arab town—Arab restaurants, bazaars, bakeries," said Rahimullah Yusufzai, a Peshawar journalist who became a leading chronicler of the Afghan war. "During the jihad, there were Arab news-papers and magazines published here. There were men in kaffiyeh, women fully covered in black."

JEDDAH

Among the first of those who answered Azzam's call was a former student of his from Jeddah, Osama Bin Laden, a young heir to a Saudi construction fortune. Bin Laden to that point had led what for well-off Saudis was a typically unfocused life. The only son of the fourth of his father's eventual eight wives, he was born in 1957, one of, all told, more than fifty children. The family business was then in the process of being enriched by its status as the favored contractor of the Saudi monarchy. Bin Laden's father, Muhammed, rose from humble beginnings—an immigrant to Saudi Arabia from the central Yemeni province of Hadramaut—to become the biggest road builder in a country that in the 1960s was suddenly flush with billions of dollars of oil wealth. Oil—a lot of it; the peninsula contains the single largest oil field ever discovered plus dozens of others—had been discovered in the late 1940s, and it would utterly transform the nation. Among other peculiar effects of the new wealth, the country was suddenly home to thousands of rich royals, with the means to acquire all the status symbols of enriched lives but little context in which to display them; they had fancy cars, but virtually no roads. Muhammed bin Laden's success building highways for those fleets of Ferraris and Cadillacs led later to the most prestigious contract imaginable in the kingdom—the restoration and reconstruction of the holy shrines at Mecca and Medina. The family's fortune was assured for generations.[11] Before that happened, however, Muhammed, who had by then divorced Osama's mother, died in a flying accident. Osama, while still a child, was allotted a share of the father's fortune. No one outside the kingdom knew how much; estimates ranged from the tens to the hundreds of millions of dollars.

Osama studied economics and engineering at university but was never granted a meaningful role in what became the Saudi Bin Laden Group. There is no consensus as to why this was so. One of the best depictions of the Bin Ladens suggested that "Osama had wanted to play a major role in the company after college but was marginalized by other brothers, either because he lacked business skills, as one source contends, or because he tried to mount an unsuccessful takeover from his elder brothers."[12]

For whatever reason, when Soviet troops crossed into Afghanistan, Bin Laden was not otherwise engaged at home. The Saudi government and religious establishment—nearly but not quite the same thing—proselytized for volunteers to assist the war effort. Critics of the House of Saud have suggested this was done mainly to appease fundamentalists at home, to demonstrate the royal family's religious commitment. The Saudi government gave wholeheartedly to the war effort, with government money, with private funds, and with volunteers. Bin Laden, always a devout Muslim, gave of his own and was encouraged to help organize some of the other funds.

To say that Bin Laden had been until then a conventionally devout Muslim might not adequately convey to Westerners the full extent of what that could mean. Saudi Arabia was one of the most socially, politically, and religiously conservative places on Earth. Its very existence was owed in large part to its commitment to that religious conservatism, which was itself the product of agreement between the Saud family and clerics of a peculiar strain of evangelical Islamic fundamentalists, known as

Wahhabis (that is, known to others as Wahhabis; they refer to themselves as Muslims). The Wahhabi sect originated in central Arabia in the eighteenth century. Its "adherents called themselves the Muwahhidun (the believers in the oneness of God). The Wahhabis had a powerful reformist message and were able to galvanize the tribes of central Arabia into a powerful military" force, according to Bernard Haykel, an expert on political Islam at New York University. The initial Wahhabi state was wiped out by the nervous rulers of the Ottoman Empire, but it reemerged in slightly altered form when King Abd al-Aziz ibn Sa'ud, commonly known as Ibn Saud, the founder of the present Saudi kingdom, based his rule and conquests on an iteration of similar doctrine known as Salafism.

The Salafis originated among intellectuals in Egypt and were later forced out of the country—this being the fate of many Egyptian dissident intellectuals. They became dominant religious influences in both Kuwait and Saudi Arabia, where the ruling Wahhabis adopted much of their ideology, which, having come out of cosmopolitan Cairo and not the desert, had a sheen of intellectual respectability that theirs lacked.[13]

Ibn Saud used military expertise and might to defeat other tribes on the peninsula. He used his religious beliefs to legitimize his victories. Those Salafist beliefs remain the all-encompassing ideology of Saudi Arabia today. "Salafism's hallmark," according to Haykel, "is a call to modern Muslims to revert to the pure Islam of the Prophet Muhammad's generation and the two generations that followed his. Muslims of this early period are referred to as al-Salaf al-Salih (the pious forefathers) whence the name Salafi. Salafism's message is utopian, its adherents seeking to transform completely the Muslim community and to ensure that Islam, as a system of belief and governance, should eventually dominate the globe. Salafis are not against technological progress nor its fruits; they do, however, abhor all innovations in belief and practice that are not anchored in their conception of the pristine Islamic age."[14]

Modern Salafis saw it as their responsibility, not a mere preference or option, to help establish Islamic rule in Afghanistan. This is the message Azzam preached, and in some cases it fell on exceptionally ready ears. Ayman al-Zawahiri, a young, politically active physician from a prominent family in Cairo, came to Peshawar as a temporary volunteer to pro-

vide medical care in 1980.[15] Zawahiri had been active in the Muslim Brotherhood in Egypt. The Brotherhood there had been waging an underground war with little success against the Egyptian government for decades. Part of its difficulties was caused by its inability to articulate a persuasive rationale for its opposition. Its critique ranged from the far right of the ideological continuum—an insistence on fundamental adherence to the Qur'an—to the far left—a demand for citizen empowerment in the government. In Afghanistan, a new rationale presented itself starkly. Zawahiri later wrote:

> The Muslim youths in Afghanistan waged the war to liberate Muslim land under purely Islamic slogans, a very vital matter, for many of the liberation battles in our Muslim world had used composite slogans, that mixed nationalism with Islam and, indeed, sometimes caused Islam to intermingle with leftist, communist slogans. This produced a schism in the thinking of the Muslim young men between their Islamic jihadist ideology that should rest on pure loyalty to God's religion, and its practical implementation. . . . In Afghanistan the picture was perfectly clear: A Muslim nation carrying out jihad under the banner of Islam, versus a foreign enemy that was an infidel aggressor backed by a corrupt, apostate regime at home. In the case of this war, the application of theory to the facts was manifestly clear. This clarity was also beneficial in refuting the ambiguities raised by many people professing to carry out Islamist work but who escaped from the arena of jihad on the pretext that there was no arena in which the distinction between Muslims and their enemies was obvious.[16]

Bin Laden, too, grasped the significance of the fight in Afghanistan and very quickly joined ranks with Azzam in Peshawar. Azzam established what he called the Office of Services, which Bin Laden underwrote. The office served as a central clearinghouse for Arab volunteers, who began to come in such numbers that somebody had to do something to sort them out and, if nothing else, give them directions to the Afghan border.

SAYYAFABAD

Peshawar in the jihad years was said to have more spies, secret agents, and freelance schemers per capita than any city in the world. Conversations dripped intrigue and purpose; there was something at stake. It was a byzantine but oddly simple world. Whatever else they wanted, most shared one overarching goal: to oust the Soviets. Everyone knew one another, prayed and socialized together, even went to the jihad training camps together. Dozens of these camps were established near Peshawar and across the border in Afghanistan by the various political parties. They tended to divide along the preexisting religious, ethnic, and political fault lines among the exile community and also by purpose— some were for basic military instruction, others for advanced weapons training, for example, or bomb making. The camps were further delineated by who sponsored each of them and to what end. Most of the camps were run by Afghans to train their own mujahideen, who were mainly peasants with no prior military expertise.[17]

The many and great divisions within the Afghan opposition required that someone attempt to unify them, or, failing that, at least identify them so that their American, Pakistani, and Saudi sponsors knew where to direct their money and expertise. Initially, the Afghans tried to organize an umbrella alliance that took in all of the opposition, but the Pakistanis refused to work with any groups that weren't expressly Islamic. The alliance subsequently was narrowed to exclude the secular opposition parties as well as any Shia Muslim groups. The man chosen to head the new alliance was a relative newcomer to Peshawar, Abdur Rasul Sayyaf, a Pashtun who had once been a junior lecturer at Kabul University and was known as "the Professor." Rare among Afghans, Sayyaf had been educated at Al Azhar University in Cairo, had spent considerable time in Saudi Arabia, and spoke fluent Arabic. He, like Azzam and Zawahiri, had been active in the Muslim Brotherhood. He received its odd amalgam of fundamentalist Islamic doctrine and modern democratic ideals selectively; the fundamentalist half of the equation was the more persuasive. As far back as the middle 1960s, Sayyaf told friends he opposed democratic reforms.[18]

At Kabul University, Sayyaf was a leader among Islamic dissidents critical of the then Communist government and was imprisoned in 1975 at the fortresslike Pol-e-Charkhi on the outskirts of Kabul, a place that became infamous for the torture that occurred within it. Sayyaf was one of the few fundamentalists in the prison; the rest had already landed in Pol-e-Charkhi's large graveyard, said Amin Farhang, a fellow inmate. The prospect of death did nothing to make Sayyaf disguise his beliefs. He still wore the long beard that was a public sign of religious conviction. Sayyaf's beard was so long, in fact, he would roll it up and put a portion of it in his jacket pocket when he ate.[19] Sayyaf and Farhang were released from prison as part of a chaotic amnesty that accompanied the Soviet invasion. Sayyaf headed immediately to Peshawar. He already knew most of the resistance leaders, some of whom had been at Kabul University with him. As the last to arrive, he was one of the few people of some stature in town who didn't already have as many enemies as friends.

"He was new, the last to arrive and thus not involved in all the internecine struggles . . . not tainted by corruption," said Yusufzai, the Pakistani journalist.

The six main resistance factions had been trying to organize themselves, at the behest of the Pakistanis, into something resembling a unified force. Sayyaf was a compromise choice by leaders of the parties to head a new alliance of all the parties.

Representing the alliance, Sayyaf traveled to Saudi Arabia, where he spoke to a conference of current and potential donors. Almost all Saudi money coming into Afghanistan had been government money delivered through the Pakistani government. There had been little private Saudi money involved. Sayyaf opened a conduit that for the next twenty years flowed in torrents. He came back to Peshawar with "a pot full of money," said Robert Eastham, U.S. consul in Peshawar at the time. "Sayyaf was a scholar, they respected that. He had fully internalized the Saudi conservative Islamic message and he could give a stem-winder, too."[20]

Not long after he returned to Peshawar, the alliance of resistance parties broke apart. Sayyaf started his own Itehad-e-Islami, or Islamic Union Party, which would become the last of seven officially sanctioned by the Pakistanis to receive aid. The Americans and Saudis agreed to the Pakistani plan.

Before Sayyaf arrived in Peshawar, Gulbuddin Hekmatyar, another member of the Afghan Muslim Brotherhood, had been the Saudi's favored warlord. Or at least that was what people thought. It wasn't something that could be proved. "Saudi money was a gravitational factor in Peshawar. You could feel it but you couldn't see it," said Eastham.[21] Sayyaf began competing directly with Hekmatyar for funding. The two of them became the favored recipients of the bountiful Saudi and, via the Pakistanis, American treasuries, amounts that would eventually total in the hundreds of millions per year. Sayyaf proved a tremendous fundraiser. He was bright and, when he wanted, could be charming. One story that made the rounds in Peshawar was that he used some of the Saudi money to fund a lavish lifestyle, but when his sponsors were due to visit, he would store all his beautiful rugs and fancy pillows, replacing them with bare mattresses. When visitors asked why he lived so simply, he would say, "How can I rest my head on a soft pillow when my mujahideen are sleeping on stones?"

"He had worked with Hekmatyar, then split. There isn't a room big enough for these two guys," said Milton Bearden, who ran the CIA base in Islamabad at the time.[22] "The only question was, Did you have real commanders lined up behind you? And he did. Of the fundamentalist parties, Sayyaf stood out because of his Arabic and his Brotherhood connections. Of the Pashtuns, he was the one."

Sayyaf's split from Hekmatyar was but the beginning of a chain of disloyalties and treachery that would mark the seven parties for decades to come, and everybody in Afghanistan would suffer for it. Nonetheless, the money made Sayyaf a figure of importance. It funded his Islamic Union Party, an army and commanders to run it, a newspaper, a huge refugee camp, and even a college called the University of Dawa al-Jihad, which means Convert and Struggle. The school became known as a place you could learn darker trades than mathematics—bomb making, for example. A student once described it to the American journalist Mary Anne Weaver as an Islamic Sandhurst, likening it to the famous British military academy, but it was also a genuine college with as many as 2,000 students studying engineering, medical technology, and literature. The school, complete with guest dormitories and athletic fields, sat behind high mud walls amid the sprawling Jalozai Refugee Camp, home to more than 200,000

Afghans, amid tobacco and sugar cane fields, near the town of Pabi, about 30 miles east of Peshawar. It was less an encampment than a city, and Pakistanis, in recognition of that fact, gave it a city's name: Sayyafabad. Locals marveled at the ingenuity of the Afghan refugees, who built out of seemingly nothing a thriving local economy that included pottery kilns, textile looms, and lumber mills churning out latticework and furniture; there were markets filled with daikon radishes, potatoes, cabbages, peppers, cauliflower, figs, dates, oranges, white melons, and tangerines. There was even a car wash.

CHAPTER 2

Those Without

FAHAHEEL

W HEN PEOPLE SAY THAT IF not for oil there would be nothing in the desert kingdoms of the Persian Gulf, it is not a great exaggeration. Before the oil, there wasn't much, but the region was not entirely empty. Before oil, there was, of course, desert—sand and the occasional date palm oasis—and few, very few, people. For a period of years after the oil, the population of the region doubled and tripled annually. It wasn't simply that oil enriched the place. It utterly transformed it.

In the case of Fahaheel, an expatriate boom town in Kuwait in the northern reaches of the Gulf, if it weren't for oil, there would not even have been a town, just a sleepy patch of farmland on the edge of an oasis on the road to Saudi Arabia. Older Kuwaitis recall that road south of the capital, Kuwait City, as a route through open country with only an occasional vehicle, camel, or Bedouin tent as landmarks. That changed with the gold-rush growth that came crashing in after the oil was discovered in the 1940s. In a decade, Fahaheel had boomed into an instant city with the jumbled cosmopolitanism of too many people brought from too many places too quickly together. The mix was similar to other places in the Gulf: Palestinians, Lebanese, Syrians, Egyptians, Jordanians, Pakistanis, Indians—even the British and Americans—came for the oil itself or, more accurately for the great majority of the migrants, for the ancillary jobs the oil economy would produce.

Mohammed Ali Doustin Baluchi joined the boom and brought his new wife, Halema, here in the early 1950s, leaving behind the hard life and harder economic prospects of his native Baluchistan.[1] Mohammed and Halema carried Pakistani passports, although, as with many Baluch, this may have been as much a matter of convenience as allegiance. Baluchistan is an arid desert territory that lies across the nominal boundaries of Iran, Afghanistan, and Pakistan, but its inhabitants often considered themselves citizens of no nation save their own. Baluchistan is one of the oldest inhabited places on Earth. It has a 1,000-mile-long coast on the Arabian Sea, and its people have gone over the water for centuries, making reputations as avid traders and warriors. Its history is one of fierce resistance to whatever would-be ruler was on hand at the moment. To cite just a few of many examples: Next-door neighbor Persia was held at bay for a thousand years; Alexander the Great suffered his greatest defeat there; and the Arab rulers who swept over the region in the century after Mohammed's death were told by their military strategists to forgo Baluchistan, for "if you send a small Army, it will be defeated, if you send a large Army, they will die of thirst and hunger."[2]

Travel within Baluch lands occurs still with little acknowledgment of the national borders crossed in the course of it. Mohammed was born in the Iranian portion of Baluchistan and Halema in Pakistan. Whatever economic promise brought them across the Persian Gulf, Mohammed found his calling as the imam of the Pakistani mosque in a Fahaheel neighborhood along the coast road known as Budu Camp.[3] It was there that the honorific "sheikh" was added to his name in recognition of his knowledge of the Qur'an and ability to teach it. It was also there that his family grew to include four sons and three daughters. The third of the sons was born in 1965 and named Khalid Sheikh Mohammed.

The site of the Budu Camp mosque, not far from the sea in southernmost Fahaheel, is now a Burger King parking lot, just a few hundred yards from the Kuwait Oil Company refinery towers at the Port of Shuaiba. The mosque may be gone; but its intentions are not: a road sign abutting the parking lot states without qualification, "Happiness in Islam."

It is through Shuaiba Port that tiny Kuwait's vast reserves—second only to those of Saudi Arabia—have been converted into riches that have made Kuwaitis among the wealthiest and most cosseted people on Earth. The traditional Bedouin birthright was a share of a father's camels; the modern Kuwaiti enjoys an endowment of guaranteed employment, free home loans, marriage bonuses, medical care, education from childhood through university, and a guaranteed income at retirement.

The main drawback to a healthy and prosperous life in Kuwait appears to be the not insignificant one that these guarantees are available to less than half the country's 2 million people. The majority of Kuwait residents are not citizens of the country and never will be. Although the government has eased citizenship requirements somewhat in recent years, most migrant workers and their offspring remain ineligible. This has created a recognizable caste system dividing those with citizenship, the native-born Kuwaitis, and those without, the guest workers, known locally as *bidoon*.

The Baluch, no matter how long they stayed, were among the *bidoon*. This was a political fact of life for the family of Mohammed Ali Doustin as well as thousands of other expatriates. In those years, Kuwait was a center of Palestinian political activism, so much so that one area of Kuwait City was known as the West Bank; the Palestinian Liberation Organization had its offices there. Yassir Arafat worked here as a civil engineer after university in Egypt. The Palestinians, in fact, predominated in the middle class and professional ranks of teachers and engineers. These included most of the teachers at the Kuwaiti schools attended by Sheikh Mohammed's children.

The boys were all very good students and technically inclined. Khalid Sheikh Mohammed's two older brothers, Zahed and Aref, attended Kuwait University, where Zahed became a student leader of the Muslim Brotherhood, the militant pan-Arab organization that functioned as an underground opposition throughout the Middle East. The campus was also home to a group called the Islamic Association of Palestinian Students, which had as members several men who would later become leaders in the Palestinian group Hamas. Zahed worked with the Palestinians and their cause became an abiding interest.[4]

During the 1970s, much of the Middle East, following the devastating Arab loss to Israel in the 1967 war and the death of Egyptian President Gamal Abdel Nasser—the erstwhile champion of a secular and united

Arab world—embarked on an inexorable turn toward religion. Secular pan-Arab nationalism had failed and religion emerged as the principal alternative source of identity and regional esteem. "It was like a huge vacuum, and nobody was able to fill this vacuum better than the rising Islamists," said Shafeeq Ghabra, a political scientist in Kuwait.[5] There was through the region a sense of terrible loss; it was as if the Arabs had lost their one true chance to regain their footing, to finally join the modern world.

Then in 1979, two seismic events shook the Islamic world: A people's army of Islamists toppled the Shah of Iran and instituted an Islamic republic; and the Soviet Union invaded Muslim Afghanistan and installed a puppet Communist government there. The two events fired imaginations throughout Islam. Politics and possibility were everywhere. Even without the pervasive media that blanket the West, ideas traveled quickly in the Arab world, largely by word of mouth. In Kuwait, conversation was abetted by the traditional style of Kuwaiti housing design in which every family home contains a room called a *diwaniya*—almost always on the ground floor, often with a separate entrance—specifically designated for discussions. They're what parlors or sitting rooms were in an older America. *Diwaniya* functioned as neighborhood club rooms. Men—and only men—gathered in them almost every evening to sit among the pillows on the floor, eat dates and drink tea, talk, watch football, discuss the events of the day. The *diwaniya* were the broadcast booths of Kuwait.

In Fahaheel, one of the prominent subjects of discussion among the expatriate Pakistanis was the long-lost dream of Baluchistan. The Baluch, like all exiles, plotted. Sheikh Mohammed's daughter, Hameda, had married Mohammed Abdul Karim, a fellow Baluch who worked in the oil industry but who refused to leave his Baluch politics at home.[6] He was a fervent Baluch nationalist, and Baluch nationalism, maybe because it had been so long unrealized, was of a ferocious kind. The warrior ethos went deep among the men of the dry plains beyond the Makran Range, the mountains that protected the Baluch interior from the sea and the enemies it carried. At various times, the armies of some Persian Gulf sheikdoms have been composed entirely of Baluch. Even Baluch lullabies reminded little boys they were in a constant struggle to gain Baluch independence and might, at any moment, be called upon to achieve it in battle. One typical Baluch folk poem advises:

Conciliation can he achieved when palms can grow hair
The jackal becomes the guard of the chickens or the fowl
Lions are grazed with the camels
Cotton becomes fireproof
Elephants are reduced to millet in size
And fish can live out of water . . .
If stone could melt away in waters then the spirit of revenge
    can be subdued
But neither can stones melt away nor can the spirit of revenge
    be extinguished in a Baluch heart
For two centuries it persists and remains smart like a young bear
    of tender age.

In Fahaheel, Abdul Karim combined this native patriotism with Salafi Muslim doctrine, a bracing combination.[7] This was the same doctrine preached by his father-in-law and fellow Baluch, Mohammed Ali Doustin.

Khalid Sheikh Mohammed was much younger than his oldest siblings and his sister Hameda had several sons near his age. They all attended high school together at a three-story brick, all-boys school that housed 1,200 students. School was, and is, a serious thing in Kuwait: schoolboys wear white shirts and gray slacks, and the headmaster walks the halls carrying a bamboo cane, to be used on obstreperous students. "Khalid excelled, especially at science," said Sheikh Ahmed Dabbous, a family friend and Islamic studies teacher at the school.

The father, Mohammed, died before Khalid graduated high school; Zahed and Aref took over the younger boy's education. They assisted in his political education, as well. As a teenager, he spent parts of his summers in youth camps in the desert at which jihad was a constant subject. Khalid became enamored with the idea.[8] He followed Zahed's lead and at age 16 joined the Muslim Brotherhood.[9] Zahed was then teaching at a local technical school. He had planned on going to graduate school in the United States, but the family decided only one boy could afford to be sent abroad—as *bidoon* they did not qualify for the generous Kuwaiti government scholarships. In a decision of enormous generosity considering what was at stake for their own futures, the older brothers chose to send Khalid to the States. They stayed behind and worked to finance his studies. "Khalid was very genius, in everything he is smart," said Dabbous. "From the beginning of his studies it's science. He wanted to go to America for this reason. He wanted to become a doctor [Ph.D.] there."

NORTH CAROLINA

Khalid enrolled at Chowan College, a tiny Baptist school in North Carolina that had recently been accepting a fair number of students from abroad. Chowan was nestled among the cotton farms, tobacco patches, and thick forests of eastern North Carolina, just south of the Virginia line. The school had been founded in 1848 as a place of learning for proper young Southern women. Later, it became a two-year junior college, a place to gain an academic foothold. Its entry standards were liberal, its values were bedrock, and its leafy setting in isolated Murfreesboro, with no bars and a single pizza shop, pretty much ensured that everyone remained on the straight and narrow. Generations of small-town ministers, teachers, and other community mainstays passed through Chowan's colonnaded façade.

After World War II, the school's missionary alumni began referring students from overseas. Dominating the international contingent by the 1980s were Middle Eastern men. Word had spread that Chowan did not require the standardized English proficiency exam then widely mandated for international students. Foreign enrollees often spent a semester or two there, improved their English and then transferred to four-year universities. Mohammed applied to Chowan as a Pakistani citizen shortly after graduating from Fahaheel Secondary in 1983. He told school administrators that he had heard of the college from a friend in Kuwait. His bill—$2,245 for the spring semester—was paid in full the day of matriculation, January 10, 1984.

Acquaintances said Khalid Sheikh was culturally integrated into Arab and Kuwaiti society and could easily have passed as a Kuwaiti. "Khalid Sheikh spoke very good Arabic, like a Kuwaiti, but introduced himself as a Pakistani," said Badawi Hindieh, a Palestinian from Fahaheel who attended Chowan at the same time. "We knew he was Baluchi."[10]

By 1984, about 50 of the 650 or so male students at Chowan were Middle Easterners, including a sizable contingent from Fahaheel and elsewhere in Kuwait. The local boys razzed them, calling them "Abbie Dahbies," a play on the name of one of the United Arab Emirates, Abu Dhabi. The Arab students were the butt of frequent jokes and harassment, especially in the anti-Muslim era that followed the 1979 Iranian takeover of the U.S. embassy in Tehran.[11]

Locals said the foreigners were cliquish. "They seemed to be praying all the time," said John Franklin Timberlake, a 1984 Chowan graduate, now a police officer in Murfreesboro. "Just chanting, like. We never understood a word of it. Sometimes we'd come home late on a weekend night, maybe after we'd had a few beers, and they'd still be praying."[12] The Arabs, in addition to their own praying, were required to attend a once-a-week Christian chapel service.

One large bloc of Middle Easterners lived in Parker Hall, a brick tower overlooking the campus's Lake Vann, a restful crescent of water frequented by migrating birds and couples holding hands. Groups of Arab students would gather in a fifth-floor dorm room and follow a kind of ritual: boil a chicken, share it with rice among all present, pray and talk, before praying again. In the Middle Eastern tradition, they would leave their shoes in the corridor. Some U.S. students could not resist the temptation to swipe the shoes, which were sometimes found floating in the lake. Another prank involved filling 55-gallon garbage containers with water and propping the big cans against the doors of the "Abbie Dahbies," then knocking and running away. When the doors opened, water flooded the rooms.

Khalid Sheikh took a preengineering curriculum popular among the foreigners; he earned good, not exceptional, marks, and left after a single semester.[13] In summer 1984, he enrolled as an engineering major at North Carolina Agricultural and Technical State University in Greensboro, a historically Black college on the Piedmont plain in the central part of the state. Unlike gentrified Chowan, A&T had an activist past. Civil rights leader Jesse Jackson is a graduate, and on February 1, 1960, students at A&T had staged the first civil rights era lunch-counter sit-in at a downtown Greensboro Woolworth's. By the time Khalid Sheikh arrived, the student body had diversified somewhat. While African Americans still made up the majority of students, there were good-sized blocs of white Southerners and Middle Eastern men.

College abroad was a rite of passage for the overwhelmingly male Middle Eastern students. Typically, this was their initial long-term exposure to Western life. Some left appalled at what they witnessed. Others ate it up.

"We were all excited about going to the States," said Khalil A. Abdullah, a 1987 A&T graduate. "In high school we had seen all the movies, heard the music. We wondered so much about it."[14]

In Greensboro, Mohammed was joined by one of his nephews, Abdul Karim Abdul Karim, who had left Kuwait at about the same time and spent his first semesters in Oklahoma before transferring to A&T to major in industrial engineering. They were part of the Middle Eastern bloc in the university's engineering department—a natural major for men from oil-producing nations.

The Middle Easterners tended to live off campus in anonymous complexes like the Yorktown and the Colonial. They seldom ate in the cafeteria and skipped organized events. While groups of "Aggies" set out to Saturday football games, the foreign students arranged soccer matches in the park.

"They hung out with one another for the most part," said Quentin Clay.[15] "The English of most of the guys was absolutely terrible. I was paired up with [Khalid Sheikh] in a senior design class. I would always get paired with one of these guys. There was much frustration. Talentwise, I questioned how they could have gotten that far."

"It was the college life: We used to get together three, four times a week, watch the games, chat, drink, you know," said Sami Zitawi, a Kuwaiti native who recalled large get-togethers of Arabs on Friday, the Muslim holy day. "We used to go to the farmers, buy a lamb or a goat, butcher it with a knife. . . . Every Friday night someone would have a big dinner: fifteen, twenty, twenty-five students."[16]

Political discussions inevitably occurred. The year before Mohammed's arrival, students in Greensboro marched in protest of the 1982 massacres of Palestinians at refugee camps in Lebanon—though the Arab visitors learned to mute their criticisms. The Middle Eastern students were far from a monolith. Differences in politics, in culture, and especially in the practice of Islam strained their solidarity.

"Basically, what you saw was a microsociety of our home," said Mahmood Zubaid, a Kuwaiti architectural engineer. "Everybody fit in where they felt most comfortable. . . . The problem is, there was no community of Arabs. There were Kuwaitis, Palestinians, Jordanians. About 200 to 300 people in total, but they tended to associate with just their own group. That's one issue. We hung around only with Kuwaitis. The community we were in, out of the two hundred or three hundred, was actually only about twenty people."[17]

An additional social barrier separated the elite scholarship boys like Zubaid from students like Mohammed: the Baluch and the Palestinians were reliant on their families or smaller grants for tuition and living expenses. But religion was the real dividing line. Wherever large concentrations of Middle Eastern students gathered on western campuses, groups of religious conservatives were established among them. These self-appointed moral overseers tried to ensure adherence to Qur'anic values and avoidance of wine, women, drugs, and other vices. They grew beards as religious statements and prayed five times a day, typically in makeshift mosques in apartments or university-provided centers. And they recruited fellow students.

"We called them the mullahs," recalled Waleed M. Qimlass, a 1980s A&T graduate. "Basically, the students at Greensboro were divided into the mullahs and the non-mullahs."[18]

Mohammed was definitely among the mullahs. Even back at Chowan, one student recalled, Mohammed had reproached him for eating pork. There was plenty at A&T for Mohammed and other true believers to be distressed about. Some Arab students drank, frequented clubs, and flirted, or more, with girls—indulging in hedonistic pursuits generally unavailable back home. A few drove around campus in Porsches and Mercedeses. The party crowd always tried to keep their indiscretions private, fearing that word might get back to their families. But the mullahs took notice and exercised pressure both intense and subtle. Islamists at Greensboro and other U.S. universities made a point of seeking out newly arrived Arab students at airports. Qimlass recalled how three "guys with beards" intercepted him and a friend, also a Kuwaiti, as they waited for their luggage at the airport in Tulsa, Oklahoma, where Qimlass studied before transferring to A&T. The trio immediately ushered the arriving students to a kind of rooming house that doubled as a mosque, reproaching a fatigued Qimlass when he lit a cigarette. If they missed new arrivals at the airport, the bearded mullahs would seek them out on campus. Their advances were sometimes rejected but often welcomed among vulnerable newcomers, who were homesick and a little bit lost in their new world.

"Your first day in Greensboro, you didn't know anybody, maybe your English is not so good, and they met you at the airport and helped you get started," Zubaid said.[19]

One Kuwaiti student wasn't so accepting: He would place a bottle of Johnnie Walker scotch on his table whenever the mullahs came by, like a cross to ward off Dracula. The disproportionate influence of religious students overseas had long troubled Arab capitals. The region's mostly autocratic rulers weren't keen to subsidize the training of would-be ayatollahs who would return and espouse Islamic revolt. Nor did the prospect of religious indoctrination abroad thrill secular parents seeking to broaden their children's horizons.

Arab governments would disperse U.S.-based scholarship students if fears emerged that any kind of religious-political cabal was gaining traction. One high official said it forever amazed him not just how many of his country's students were changed, but which ones. "We had a lot of our students coming back from the U.S. radicalized. I'm not talking about religious guys going to the U.S. and coming back as fundamentalists. I'm talking about cool guys," he said.[20]

"Why would they flip religiously? It happens there," said another Kuwaiti student.[21] "When we are there we are very vulnerable. That's why we get into groups—to protect each other. The religious guys work on them. Why is it so easy? The key thing I think is the political views more than the religious."

Students who recalled Mohammed invariably described a studious and private devotee of the library and Allah, but friendly enough in a casual way and capable of a laugh. "All anyone knows about him is that he was in the mosque all the time," said Faisal Munifi, who studied mechanical engineering.[22]

"He very much kept to himself," said Zitawi. "We'd see each other at the Burger King for coffee or lunch. That was our hangout. . . . He was always polite. He wasn't a funny guy, but when he's talking to you, you feel like he's smiling. He wasn't rude or anything."[23]

Nor did Mohammed spout anti-Western or anti-American rhetoric. "Something must have happened later that caused that feeling," said Hindieh, who knew Mohammed at both Chowan and Greensboro. "I never remember him saying anything like that."[24]

Whether he said anything or not, something had happened. Khalid Sheikh's three brothers—Aref, Zahed, and Abed—had moved to Peshawar, Pakistan, to support the jihad in Afghanistan. On at least one summer vacation, he joined them. "These trips affected him," said a Kuwaiti security official.[25] So had his stay in the United States.

"When he goes there, he sees most Americans don't like Arabs and Islam," said Dabbous, his high school teacher:

" 'Why?' I ask him.

" 'Because of Israel,' he says. 'Most Americans hate Arabs because of this.' He's a very normal boy before. Kind, generous, always the smiling kind. After he came back, he's a different man. He's very sad. He doesn't speak. He just sits there.

" 'Why?' I ask.

" 'Because of what I am saying about the Americans hating Islam,' he said.

"I talked to him, to change his mind, to tell him this is just a few Americans. He refused to speak to me about it again. He was set. This was when he was on vacation from school. When Khalid said this I told him we must meet again.

"He said, 'No, my ideas are very strong. Don't talk with me again about this matter.' "[26]

By the end of 1986, after just two and a half years, Mohammed had completed his school work. He and his nephew Abdul Karim were graduated December 18. Khalid Sheikh was one of twenty-eight mechanical engineering graduates, almost a third of them Middle Easterners. As at Chowan, there was no photo of him in the yearbook.

JALALABAD

The original jihadis started in old Peshawar with very little money, in the pre-Saudi, pre-CIA days," said Yusufzai, the Pakistani journalist. "Later, they all rented places in University Town, the most expensive neighborhood in Peshawar."

Those University Town neighborhoods were, oddly, the most Westernized in the city. Old Peshawar was a crooked tangle of alleys and bazaars thick with the smells, sounds, and people of Central Asia. A dense haze of exhaust, dust, and brick kiln smoke lay over it. It had the feel of a place where history was not only made, but present. University Town, by contrast, was clean and rectangular, laid out on a grid filled with walled compounds of big three-story stucco houses that would be at home in any California subdivision and had about as long a lineage. The new villas were filled by the Arab fighters and an even larger militia of camp followers.

Armies used to be trailed by merchants of flesh, food, and other entertainments; modern armies, even ragtag agglomerations like the mujahideen, are as likely to be followed by social workers as streetwalkers. The Afghan wars, because of the international nature of their combatants and finances, were the apotheosis of this. After the arms trade, the biggest industry in Peshawar during the war was good works. More than 150 charities, development, and refugee care organizations opened offices. They were about evenly split between Western and Arab world sponsors. There were so many charities they required three separate coordinating councils to sort out their work. There was plenty to do. Afghanistan at the time of the Russian invasion had a population of 15 million. Over a decade, that would shrink by almost half. The missing millions had to go somewhere and most fled through the mountain passes to Pakistan. The Pakistan-Afghanistan border in theory has only a handful of official crossings. In fact, there were hundreds, so many the border was never much more than a line on maps.

One of the largest aid agencies was a Kuwaiti charity called Lajnat al Dawa al Islamia, the Committee for Islamic Appeal. LDI, as it was called, at one point had more than 1,200 employees in Pakistan and was spending $4 million a year in the region.[27] Among other efforts, LDI ran a 200-bed hospital in Peshawar and another inside Afghanistan, as well as numerous smaller clinics. It also funded and staffed twenty-two Qur'anic study centers. Its regional manager, beginning in 1985, was Zahed Sheikh Mohammed, Khalid Sheikh Mohammed's older brother. Zahed had left his teaching position and gone to work for LDI in Kuwait in 1983 and was posted to Peshawar two years later. As head of one of the largest charities in town, Zahed became a figure of importance. He knew local diplomats, the Afghan warlords; when Pakistani politicians came to town, he shared the dais with them.[28] Zahed worked out of an office on Arbat Road in University Town. Many of the Arab charities were suspected of being little more than fronts for the distribution of jihad money, but Zahed Sheikh's LDI was not thought to be among them. Whatever, if any, assistance it provided to resistance fighters, it demonstrably spent millions of dollars on schools and clinics and the promotion of the particularly Kuwaiti brand of Salafi Islam. Other Arab charities did the same, and religious schools flourished not just in Peshawar but throughout Pakistan and in areas of Afghanistan controlled by the resistance.

After finishing college in North Carolina, Khalid, according to Kuwaiti authorities, never returned home to stay. Instead, carrying a freshly minted dislike of the United States, he joined his big brother in Peshawar. Another brother, Abed, a secondary schoolteacher and Islamic scholar, left his job in the Gulf emirate of Qatar and came east, too. A man who knew all three said Zahed, the eldest, was the coolest head of the trio; Abed was the more militant and Khalid tended to be more like him.[29] All three wore trim black beards and, typically, checked headscarves. Zahed and Khalid were somewhat short in stature and slightly overweight. The two looked enough alike that, despite their age difference, some people could not tell them apart.

Abed worked for Abdur Rasul Sayyaf's newspaper in Peshawar and

Khalid taught engineering at the warlord's university, a friend said.[30] Khalid also worked at the huge Jalozai refugee camp. Sayyaf attracted men from all over the Muslim world, not just Arabs. Southeast Asian Muslims who came to the jihad trained almost exclusively with him. Khalid Sheikh made the acquaintance of several southeast Asians, including an Indonesian, known by the nom de guerre Hambali, who would remain a friend for the next fifteen years.

The three Baluch brothers settled in. Khalid married a woman he had met at Jalozai. They became part of the small, semipermanent Arab community that included Azzam, Islamic Jihad founder Ayman al-Zawahiri, and Bin Laden, who came and went with his wives and children in his own airplane. "It was easy to see Sheik Osama, who lived openly in the Hayatabad part of Peshawar," said one man who was in the movement.[31] "He had been visiting Peshawar on and off during the mid-1980s when he was also spending a lot of time across the border, fighting in Afghanistan. Zawahiri came [to stay] in 1985. Bin Laden then moved his wives and children to Peshawar in 1986 and lived there until October 1989. During that time, he visited Saudi several times a year."

Abed and Khalid worked closely with Azzam; Khalid oversaw distribution of relief and military supplies.[32] He struck others as bright, quite clever and knowledgeable about the broader politics and implications of the war.[33] The brothers sat in on meetings with party leaders Burhanuddin Rabbani, Gulbuddin Hekmatyar, and Sayyaf. Except for Bin Laden (who already had begun to regard himself as in some way apart), most of the Arabs prayed at the small Saba-e-Leil (Lion of the Night) Mosque in a bustling neighborhood on a dead-end alley off Arbat Road, not far from both Sayyaf's and Zahed's offices. The neighborhood had bakeries, halal butchers, tailors, and, of course, travel agents for booking passages home. One reason Peshawar took on so much of an Arab effect was that most of the Arabs who came for the war never got far beyond the city; most never actually saw combat. And when they did it was to minimal effect. They were largely regarded by the various Afghan armies as a necessary nuisance to be abided in exchange for the money that followed them. Burhanuddin Rabbani, one of the resistance leaders, said: "I was against the Arab volunteers to come and fight with us. When they came to me, I told them they shouldn't have come, they should have saved the money for the ticket and instead donated the money for our struggle. The Arabs were not experienced fighters, they came from the desert, they didn't know the mountains, they weren't familiar with our traditions and culture."[34]

Ahmed Wali Massoud, brother of the Northern Alliance commander Ahmed Shah Massoud, agreed: "I don't know if the Arab fighters were good or bad fighters, no one saw them. There were so few."[35]

Some Afghans found the desert Arab version of Islam strange and resented being lectured on the proper way to worship. They perceived the Arabs as haughty.

"Saudi, Gulf families would send their kids to Afghanistan in the summer for some school project," Eastham, the American consul in Peshawar said. "They'd send them through a training camp, give them a weapon and send them into Afghanistan. They worried to death about it, didn't want them to get killed. These were the donors' kids. You had to accept them. It drove the Afghans crazy because the Arabs had an air of superiority about them, an arrogance."[36]

The lack of opportunities for the Arabs in the field was a point of prolonged and heated debate among them. The way Afghan commanders employed the Arabs—or, more to the point, that they often refused to—rankled. Many Arabs, among them Bin Laden, agitated for formation of separate Arab fighting units. Azzam, who held sway over the community, insisted the Arabs must do what the Afghans asked; it was their war, not the Arabs'. Azzam maintained that until the cause was complete in Afghanistan, until the infidels were driven away, his Office of Services would do just that—serve. But resentments accumulated and Bin Laden began to branch out on his own. In his first years in Peshawar, he had functioned solely as a financial backer and fund-raiser. By 1985, he had brought in heavy construction equipment from Saudi Arabia and begun building training camps, roads, and runways across the border in Afghanistan.[37] In 1987, he and a small group of Arab fighters, including Azzam, Khalid Sheikh Mohammed, and their Afghan patron, Sayyaf, took part in a battle with Russian troops attempting to take the village of Jaji.[38] The Arabs held out for weeks. The Soviets eventually abandoned the fight, and it was this victory that formed the foundation for the legend of Bin Laden as a fierce battlefield warrior.[39] Bin Laden certainly saw it that way. He told stories about the Battle of the Lion's Den, as he sometimes called it, for more than a decade after.[40] Arab journalists reported on the battle extensively, highlighting Bin Laden's role. Many of the men who fought at Jaji would later take large roles in Al Qaeda. After the battle, Bin Laden came to believe that there was a goal beyond Afghanistan, a global jihad to be fought and that he was fated to lead it. He began to position himself for the task.[41]

In the end, at least in so far as Afghanistan was concerned, Arab participation in the war didn't matter greatly. The Afghan resistance, heavily armed and funded by its backers—which grew to include China, Japan, Great Britain, and Israel in addition to the United States, Saudi Arabia, and Pakistan—proved more than able to bedevil the Soviet troops all on their own. As Brzezinski had hoped, the Soviets sank deeper and deeper into what became their version of the American Vietnam nightmare. Rather than a quick and decisive engagement, the campaign drew on year after year, the rolls of the dead and injured grew, and criticism at home mounted. The expense of the war was huge as well, and the Soviet Union, unbeknown to its enemies in the West, was frighteningly near economic collapse. When Mikhail Gorbachev came to power in 1985, he concluded quickly that the war was unwinnable. Even before Gorbachev, the Soviets had made approaches toward a negotiated end to the war, but those all came to naught, in large part because the Soviets wanted to leave a government friendly to them in power. The Pakistanis, in particular, insisted throughout the war that any future Afghanistan government had to include the fundamentalist parties, which they saw as compliant to their own goals. In effect, the Pakistanis intended to take control of Afghanistan's future. Rather than enter negotiations with the Soviets, the patrons of the resistance increased their assistance in both amount and kind. In 1985, for example, the United States began supplying handheld Stinger antiaircraft missiles to the mujahideen, the first time those sophisticated weapons had been provided outside traditional American alliances.[42] Already by the end of 1986, unknown to the resistance, Gorbachev had decided he would leave Afghanistan with or without victory and in 1988 he started actual troop withdrawals.[43] By February 1989, the last Soviet soldier had left. The Soviet adventure in Afghanistan was over, and with it an era in world history was about to close.

Having failed to negotiate an exit,[44] the Soviets left behind a government, led by the former head of the secret police, Mohammed Najibullah, friendly to their interests. So the war didn't stop, or even slow down, becoming instead a war of Afghan against Afghan, albeit with significant assistance from without. The Najibullah government was sufficiently weak that certain elements within the resistance thought it could be knocked out with a single significant blow. The plan, pushed by Sayyaf and strongly favored by the Pakistani ISI, was for a direct attack on the government garrison town of Jalalabad in southeast Afghanistan. It was the Afghan city closest to Peshawar and thus a natural target for an offensive; the resistance had the whole of Pakistan as a marshaling yard for the attacking forces.

"The idea of Jalalabad was [to win] a big psychological victory," said Robert Oakley, then the U.S. ambassador to Pakistan. The offensive was backed by the new Pakistani prime minister, Benazir Bhutto. Bhutto, in fact, attended the key meeting in Islamabad where the plan was put in motion. "She went along with whatever the ISI proposed. She had no choice and she wanted to show that she was macho. Sayyaf was the one really pushing it. It was his area, his deal. We didn't try to stop it. We thought it might work, too."[45]

Milt Bearden, the CIA station chief in Pakistan, had a less optimistic view of the plan's prospects. It was his opinion that Bhutto was the primary proponent, wanting a victory she could take to the international community to strengthen her bargaining position. "Bhutto pushed ISI," Bearden said. "We didn't need it. I didn't need it. There were always other interested parties. It was a real goat fuck. Brigadier Yousef [of the Pakistani ISI], he was calling us, asking us why we were doing this, 'It's a bad idea.' Honestly, the CIA would not have done it that way, a fixed battle. We would have tried to negotiate something at that point. After February 15 [and the Soviet withdrawals], I would have been much more inclined to deal making."[46]

No one else was much inclined to cut deals. The general feeling among the American and Pakistani overseers was that the Najibullah government would likely collapse of its own weight in short order, especially if it were given a shove. The battle of Jalalabad was launched in the spring. Its intent was to take control of the city and the surrounding region, declare it the provisional capital, and install an interim government made up of resistance leaders.[47] Jalalabad was within easy reach of Peshawar. The Pakistanis thought they could just drive the new government up and drop them off and the government in Kabul would be swept away in the resulting momentum. Several thousand Afghan mujahideen mustered for the assault. In a rare bit of inclusion, several hundred Arab fighters went with them. It would be one of the largest fixed engagements of the war, and, for the resistance, one of the biggest disasters. The initial thrust into the city carried the mujahideen to within sight of the airport, but it stalled. They laid siege to the city, thinking it couldn't possibly hold out for more than a week. The siege lasted for two months. The opposing armies rained rockets down on their enemies, killing thousands.[48] As with much of the war, little regard was given innocent civilians caught in the cross-fire. They died by the hundreds. Thousands of refugees trudged out of the city toward Pakistan. A Pakistani general later wrote the attack was a failure of planning as well as execution: "The U.S. shipments were still substantially less than necessary, the reserve stocks had never been built up again after the Ojhri Camp disaster, and there had been little forward planning or dumping of available stocks prior to the battle. Not only was the strategic wisdom of attacking Jalalabad doubtful, but the tactics and logistics of carrying it out were quickly revealed as inadequate. . . . ISI and the Party Leaders made a strategic blunder in moving from guerrilla to full-scale conventional warfare too soon. . . . Tactically, it was a textbook example of how not to fight a battle."[49]

The Arab fighters suffered large casualties, too. Among the dead was Abed Sheikh Mohammed, the youngest of the Baluch brothers from Kuwait. Azzam, in typically florid style, eulogized him:

"The souls of the best propagators had been taken to Allah, and the best of the people passed on to Him in this month, and Jalalabad was left crying, 'Is there any more fuel for the battle of Iman? O Nangarhar [the province in which the battle was fought], has your thirst not been quenched for the blood of the pious? Do the graves of the pious which you have swallowed not suffice you? Enough! Enough! You have taken the children of our hearts and souls.' . . . [Abed's] soul refused that he live a life of luxury under the air-conditioners while his brothers in faith were being killed under the gunfire of Allah's enemies, and their children were dying in the summer heat of Peshawar, at an average of one hundred a day. How can a lion accept to live chained up in a zoo, put on a show for all to see? He must break the chains and return to his den, with the rest of his fellow lions."[50]

After Jalalabad, the various Afghan factions, deprived of a common enemy, began fighting one another as well as the Kabul government. Hekmatyar, who was still a major recipient of U.S. and Pakistani funds, was accused of systematically murdering rivals and rebellious commanders. Open war broke out between Massoud and him. It marked the beginning of a descent into internecine conflict that would engulf the country for the next decade. As the Afghans turned on one another, the Afghan Arabs in Peshawar engaged in their own debate about the future of their holy war. The Soviets were gone. If ridding the country of the Communists was the noble calling, should the campaign now end? Many of the Arabs, including Bin Laden, argued that their duty now was to take the jihad home, to take the lessons learned from Afghanistan and insist that their own countries be ruled by Islamic law. Azzam argued they needed to stay the course, to focus all their resources and energies to see an Islamic government installed in Afghanistan; until that was achieved, there was no victory. He tried to negotiate a peace between Massoud and Hekmatyar toward that end, but like everyone who had ever tried to negotiate an end to Afghan fighting, he failed.

A similar debate was occurring within the American intelligence and diplomatic bureaucracies over the continued U.S. role in Afghanistan: With the Soviets gone, who should the United States support?[51] That debate was never resolved. In the near term, different portions of the U.S. bureaucracies supported different strategies in Afghanistan, in effect supporting everyone.[52] In the long term, the United States decided it was better off supporting no one and largely left the Afghans to their own (and the Pakistanis') devices. Many resistance leaders felt that the United States after the Soviet withdrawal actively opposed the establishment of an Islamic government in Kabul. This was, to some, the cruelest cut. When the Americans left, they didn't even say "bye-bye, Afghans," said Ahmed Shah Ahmedzai, a Sayyaf deputy.

Late that year, Azzam, the heart—the creator, really—of the Arab jihadi resistance, and two of his sons were murdered by a remote-control bomb on Peshawar's Arbat Road, just outside the Lion of the Night Mosque. They were en route to Friday prayers. The murder was never solved. The political and religious climate began to change in Peshawar, and resentment of the American abandonment festered. Bin Laden took control of the Office of Services and, at least in the popular imagination as well as his own, he effectively replaced Azzam as head of the Arab mujahideen. Even before Azzam's death, he had begun recruiting men to join a new group to carry the jihad beyond Afghanistan.[53] He called the group Al Qaeda, The Base.[54] Then came the 1990 Iraqi invasion of Kuwait and the American-led counterattack, which deepened divisions within the Arab world. Bin Laden, for one, was furious that the Saudi royal family allowed the United States to base its soldiers in the kingdom, violating what he felt was a dictate to keep infidels out of the holy land.[55] He confronted the royals, offering to raise his own holy army to protect the kingdom. The family, flabbergasted at his naiveté, declined the offer.[56] He attacked the U.S. and the Saudi governments in speeches, saying there was no difference between the Soviets and the Americans—they were all infidels and would meet the same fate, to be swept away in the coming Islamic tide.

CHAPTER 3

World War

T
NEW YORK

HE SECOND PHASE OF THE AFGHAN WAR was as bloody and, if possible, even crueler than the first. The fact that it was now Afghans fighting Afghans dimmed enthusiasm for the war in the imperial capitals, but it did little to lessen its attraction within the Muslim world. Jihad enthusiasms had been invigorated by the defeat of the Soviet Union. If anything, the magnetic attraction of holy war grew. Many men who had come to fight the Soviets became missionaries for the jihad cause. They returned home flushed with glory even if the Arab volunteers had done little to win the war. They bore witness to the power of Allah on the battlefield. It was the power of the Prophet revisited, reborn in the here and now. The jihad veterans sought to awaken their coreligionists to the Muslim history of conquest. In the most grandiose imaginings, they urged a new, broader holy war to reestablish the Islamic caliphate—a Muslim empire—no matter what that implied. More practically, the men who had come to Peshawar, who had trained and fought, or not, met like-minded men from around the world. And not only Arabs. There were hundreds of jihadis from East and Central Asia, from North America, from Europe, from wherever Muslims lived, which would include nearly every place on Earth. No one has an accurate count on the number of volunteers to the Afghan cause. Estimates range from 25,000 foreign fighters to triple that during the Soviet war, and the caravan did not slow in the least after the

Soviet retreat. The war became, instead, what one terrorism expert called a conveyor belt, training then disgorging a steady supply of men ready to fight wars elsewhere that didn't yet exist.[1] Or, as it turned out, create them. The veterans of the Soviet war formed an informal network that recruited, encouraged, and assisted others to do as they had done. A global web was woven and waiting.

The established training camps flourished and new ones were built as various warlords, armies, and other agencies perceived a need for highly motivated young men who knew how to handle small arms and explosives. With the unending war, the training camps, the madrassas, and the related logistical needs, Peshawar was as hectic as ever. Among the men who came through after the war was Khalid Sheikh Mohammed's nephew, Abdul Basit Abdul Karim. Abdul Basit, the brother of Khalid's college classmate, Abdul Karim, had grown up in the same Kuwait suburb as Khalid and was just three years younger. He had left Kuwait for university, choosing a small technical school in Wales. While he was away, his family left Kuwait and returned to their native Baluchistan.[2] The main livelihood in the region was war work—smuggling, contracting, and trading military goods. Basit's father and two of his brothers took active roles assisting the Afghan resistance.[3] Abdul Basit had first come to Peshawar, like thousands of others, on a summer holiday from college, in 1988. After receiving a technical degree in electrical engineering in 1989, Basit returned to Kuwait and briefly worked for the Kuwaiti government,[4] but that and all other normal life in Kuwait came to an abrupt halt with the Iraqi invasion in August 1990. It is uncertain exactly what happened to Basit during the invasion, but by the spring of 1991 he was back in Pakistan. In March in Chakiwara, a dense Baluch neighborhood of Karachi, he married Latifa Abdul Aziz.[5] Basit paid a dowry of 10,000 rupees (about $400). After the wedding, he returned to Peshawar, where, he later told investigators, he spent approximately six months in the training camps.[6] Other men who attended the camps in that period said Basit also worked as an instructor.[7] He had studied electrical and computer engineering in college and used some of that knowledge to develop an expertise in bomb building; he became known to trainees as the Chemist.[8]

In addition to meeting up with his uncles, Abdul Basit renewed acquaintances with jihadis from Southeast Asia he had met when he first came to camps three years earlier. The different training camps tended to develop informal supply lines for new recruits; like universities in the United States, later-arriving students tended to go where their national predecessors had gone. Abdur Rasul Sayyaf's camps from the beginning had been the principal destination for recruits from Muslim Southeast Asia—Indonesia, Malaysia, and the southern Philippines.[9] These men made friendships at the camps that would last for decades, both among themselves and with Afghan and Arab jihadis. One group of Filipinos returned home and founded an Islamic resistance group there, naming it Abu Sayyaf after their Afghan patron. The leader of Abu Sayyaf, Abdurajak Janjalani, met Abdul Basit at the camps and invited him to Basilan Island in the Philippines to teach at his new guerrilla camp.[10] Abdul Basit made an initial trip to the Philippines sometime in 1991. He wasn't impressed. Abu Sayyaf was a ragtag organization. Its stated goal was to create an Islamic state in the southern Philippines, but its operations tended more toward the purely criminal than the political. Its real expertise eventually evolved into kidnap-for-ransom schemes that were ends in themselves. Abdul Basit later told a friend it would be difficult for him to help the group as their "main interest is in the use of firearms," not Basit's expertise.[11]

By the end of the year, many of the original Afghan Arabs had left Peshawar, including Bin Laden who had returned to Saudi Arabia. Before leaving, he had formed his fledgling Al Qaeda organization to coordinate jihad activities beyond Afghanistan. He had already sent people to Sudan to set up an operations base there,[12] and, from his home in Jeddah, he tried to organize an insurgent army to fight the Marxist government in South Yemen.[13]

Those who stayed behind in Pakistan changed perceptibly. The circle that included Basit's uncles, Khalid Sheikh and Zahed Sheikh Mohammed, moved with the times. The Soviets had left. Azzam and Abed were dead. One friend of the Baluch brothers said: "In 1991–92, their whereabouts, their meetings, their thoughts, it became more secret. The hatred for Americans—it was among every Arab who came to Afghanistan. . . . They all thought America had imposed its rule by vetoing Islamic rule in Afghanistan through its agents in Pakistan."[14] Many thought they had been duped into serving U.S. interests against the Soviet Union, used and then tossed aside.

While many mujahideen who had gathered in Pakistan went home, warriors without a war, others went searching for a new one. Some, including Khalid Sheikh, found it, at least briefly, in Bosnia.[15] Khalid Sheikh went to the Balkans in 1992, but didn't stay long. He returned to Pakistan, then, with the sponsorship of a member of the ruling family of Qatar, moved his family to Doha, its capital, where he took a job as an engineer in the tiny sheikdom's public works ministry.

Basit returned briefly to Peshawar too, but he was already set on his own course. It has never been clear which cause, if any, Basit was serving. Given the haphazardness of his activities, and their early lack of sufficient funding, it seems most likely he was a freelancer, the harbinger of a new type of independent, non-state-sponsored global terrorist. Over the next year, he began to make plans to attack the United States. He contacted Abdul Hakim Murad, a boyhood friend who was then in the United States studying at flight schools. Basit told him he wanted to attack Israel, but thought it was too tough a target, so he had decided to attack the United States instead. He asked Murad to suggest potential Jewish targets in the United States. Murad agreed to think about it. Later, after Murad finished his training and returned to the Gulf in 1992, Basit contacted him again:

"And he asked me if I selected a place. . . . I told him the World Trade Center. He asked me why and I gave him the reasons. I asked him what he was going to do. He told me that he took training for six months in Afghanistan. I asked him what kind of training. So he told me, 'Chocolate.' I answered, 'What do you mean by chocolate?' He said, 'Boom.' And I immediately understood that he took training in explosives and he told me it is time to go to the United States."[16]

Why Jews? Why the United States? Murad said:

"I was working for my religion because I feel that my Muslim brothers in Palestine are suffering. Muslims in Bosnia are suffering, everywhere they are suffering. And if you check the reason for the suffering, you will find that the U.S. is the reason for this. If you ask anybody, even if you ask children, they will tell you that the U.S. is supporting Israel and Israel is killing our Muslim brothers in Palestine. The United States is acting like a terrorist, but nobody can see that. I mean supporting Israel by money and by weapons, that is considered also a kind of terrorist."[17]

With little more preparation than Murad's suggestion that a lot of Jews worked in the twin towers, Basit went to a travel agency in Peshawar and purchased, at a cost of $2,500, two one-way first-class tickets from Peshawar via Karachi to JFK International, one for himself and one for a man he had met in Pakistan, a hapless Palestinian named Ahmad Mohammed Ajaj.[18] Ajaj had emigrated to the United States, to Houston, Texas, and had applied for political asylum, claiming he had been persecuted by the Israeli Army in the occupied West Bank. While he was waiting for that claim to be heard, he worked as a watch salesman and a Domino pizza deliveryman. Then, he said, when the chance presented itself—an acquaintance offered him a good price on an airline ticket—Ajaj traveled to Pakistan to see the jihad firsthand.[19] He neglected, however, to obtain the necessary documents to ensure his reentry to the United States. Abdul Basit offered him a return ticket and he took it. When they arrived at JFK on the evening of September 1, 1992, they were both stopped at customs. Ajaj was using a crudely forged Swedish passport Basit had given him, and he was stopped at entry. When he admitted the passport was not his, he was detained. A search of his three pieces of luggage revealed very few articles of clothing but several military manuals, jihad magazines, recipes for chemical explosives, and more false IDs. He was jailed and he stayed in custody for the next six months.

Basit, on the other hand, was traveling light. Most of what was in Ajaj's luggage apparently belonged to him and he arrived with only a single carry-on bag. He was dressed like a peacock, one man said—in a bright orange, brown, and olive green three-piece silk ensemble with flowing sleeves and ballooning harem pants. He carried what appeared to be a valid Iraqi passport in the name of Ramzi Ahmed Yousef. But the passport contained no entry visa for the United States.[20] When asked about the lack of a visa, he told the customs agent he was fleeing political persecution in Iraq, which had been in a nearly constant state of war with the United States since the Kuwait invasion two years prior. He was given a choice of being deported or arrested. He chose arrest and was immediately freed on his own recognizance, walked out of the airport and took a taxi to Brooklyn. Thus began the career of Ramzi Yousef, master terrorist.

Yousef went directly to a Brooklyn mosque that housed the branch location of Azzam's Office of Services. He met the spiritual leader of local radical Muslims, Omar Abdel Rahman, an Egyptian cleric known as the Blind Sheikh. Rahman had been a confederate of Azzam's and was one of the chief theoreticians of modern Islamist movements around the globe. Through Rahman and others at the mosque Yousef was introduced to a variety of young activists. He quickly assembled the crew of men who would assist him in his fledgling campaign against the United States. Yousef, as a prospective terrorist, had two great abilities: his technical knowledge of explosives and his charm. He was a tall, lanky man, saved from handsomeness by outsized ears, but with an almost raffish quality. He was, it seemed, able to persuade people to do things for him that they had either never thought of or were unwilling to do on their own. It seemed effortless. He was skilled at choosing assistants who were often not especially bright or sophisticated, men of slight abilities and accomplishments, who he was able to order about as he saw fit. As his old friend Murad explained, Yousef wasn't a particularly religious man—or perhaps not religious at all—but he was able to speak the language of religious men and make their goals seem akin to his own.

In the fall of that year, 1992, Yousef recruited a team of largely marginal, unaccomplished men to join him in building a bomb. He moved into an apartment in Jersey City, New Jersey, and in between assembling the team and the bomb, scouted targets. He drove around Brooklyn because he had been told Jews lived there. He considered several different types of bombs. He dreamed of building a device in which the initial explosion would release a cyanide cloud into the World Trade Center's ventilation system, poisoning everyone within. He even bought some cyanide to test it, but he lacked the money to build that device. Eventually, he settled on the most inexpensive model he could conceive, a basic fertilizer bomb.[21] He periodically took prototypes of the bomb out into the New Jersey countryside and blew them up. In February, after at least one false start that ended in the hospital after an auto accident, a bomb Yousef built for

$3,000 blew up in the basement of the World Trade Center's North Tower, killing six people, injuring 1,000, and causing $300 million in damage. He and his compatriots stopped at a mailbox in Manhattan on their way out of the city. They mailed letters to five news organizations, claiming responsibility. At one point later in his life, Yousef described himself as Pakistani by birth, Palestinian by choice. The letter received by the *New York Times* reflected this:

We are, the fifth battalion in the LIBERATION ARMY, declare our responsibility for the explosion on the mentioned building. This action was done in response for the American political, economical, and military support to Israel the state of terrorism and to the rest of the dictator countries in the region.

OUR DEMANDS ARE:
1. Stop all military, economical, and political aid to Israel.
2. All diplomatic relations with Israel must stop.
3. Not to interfere with any of the Middle East countries interior affairs.

If our demands are not met, all of our functional groups in the army will continue to execute our missions against the military and civilian targets in and out the United States. For your own information, our army has more than hundred and fifty suicidal soldiers ready to go ahead. The terrorism that Israel practices (Which is supported by America) must be faced with a similar one. The dictatorship and terrorism (also supported by America) that some countries are practicing against their own people must also be faced with terrorism.

The American people must know, that their civilians who got killed are not better than those who are getting killed by the American weapons and support.

The American people are responsible for the actions of their government and they must question all of the crimes that their government is committing against other people. Or they—

Americans—will be the targets of our operations that could diminish them. We invite all of the people from all countries and all of the revolutionaries in the world to participate in this action with us to accomplish our just goals.

Liberation Army Fifth Battalion

Al-Farrek Al-Rokn, Abu Bakr Al-Makee

There was no Al-Farrek or Abu Bakr. There was no army, either— just Yousef and his fervid imagination, which was quite enough. As Yousef sat that evening in Jersey City, looking at the Manhattan skyline, he was disappointed. The news reports carried estimates of the damage and death. The bomb did less than he intended. Yousef originally wanted the bomb to topple the North Tower onto the South Tower, knocking down the entire complex. He estimated this would kill 250,000 people, although where he came up with the number is anyone's guess; it is many times more than the number who actually worked in the buildings.[22] Yousef drafted another version of the letter claiming responsibility, acknowledging the bomb had been somewhat less damaging than he hoped. In that letter, which was read at trial but never sent, he issued a warning: "Unfortunately our calculations were not very accurate this time; however, we promise you that next time, it will be very precise and WTC will continue to be one of our targets unless our demands have been met."[23]

Almost everything about the attack on the World Trade Center seemed remarkably casual, so much so that a CIA official later termed Yousef an ad hoc terrorist. Yousef appeared to be an impetuous fellow, bright enough and creative enough, but not in the least concerned with fine details. He had, for example, recruited the man who drove the World Trade Center bomb to the towers the night before the attack. He had originally intended his trip to the United States as more of a reconnaissance mission than an attack. One result of this was that he was woefully underfunded. He spent more on his air travel in and out of the country than he spent on the bomb itself. He said he determined the date of the attack by the heft of his wallet. When he ran out of money, be blew up the bomb. He said the money came from family and friends, but would not specify further. The only confirmed source of any of his funding was a $660 wire transfer his uncle, Khalid Sheikh, sent to him in New Jersey. If he had had more money, he said, he would have built a bigger bomb.

But the end result of Yousef's attack—six deaths and hundreds of millions of dollars in damages—was certainly not insignificant. More than the physical effects, Yousef demonstrated that even with his group's ineptitude, he was able to strike successfully at the heart on his enemy. The World Trade Center was not an inconsiderable target. His approach was haphazard, his methods amateurish, but his intent was real and he demonstrated beyond doubt that dire results were possible. The question that should have been asked after the first World Trade Center attack was not: How was he able to do that? but, How lucky were we to have dodged a much larger disaster? With very few changes, Yousef could have killed thousands.

To a considerable extent, America did not recognize the advent of a new age but whether anyone knew it or not, an era of religious terror had arrived. Intermingling religious and political goals had been the norm for most of human history. Islam itself came into the world with secular as well as sacred aims. What had changed in this latest incarnation had more to do with the world it was in than Islam itself. By the latter half of the twentieth century, the movement toward secular government had triumphed almost everywhere except in the Islamic world. The advocates of political Islam became aberrant simply by outlasting the political ambitions and empires of other religions. They might have been mere curious anachronisms had not the modern world provided them the means to wed their old beliefs to new, readily accessible technologies. The outcome of that union is terror on a scale not previously known.

KARACHI

Yousef flew out of the country the night of the attack, using his real name, Abdul Basit. Others weren't so fortunate. Yousef had done nothing to take care of his crew. He left them all to their own devices. In the end, that was a mistake. The bomb had been carried into the World Trade Center garage in a rented Ryder van. The man who arranged the rental, Mohammed Salameh, had no money with which to arrange his own escape afterward. In order to buy a ticket to flee he tried repeatedly to get money from the only place he knew he had a decent chance of succeeding: by reclaiming the $400 deposit he had made on the rental van, which he told the rental agency had been stolen. Unbeknown to him, bits and pieces of the van survived the blast. Investigators recovered the vehicle identification number and traced it to the rental agency. From there, it was child's play to find and arrest Salameh, who, in his desperate attempt to get the money, delivered himself to police.

His arrest and others that followed built a trail directly to Yousef, who had fled—as usual, in first class—back to Pakistan. An international manhunt followed, with a reward of $2 million for his capture. The U.S. government printed 32,000 matchbooks with Yousef's photo on them and air-dropped them into the country. Yousef disappeared for a time in Baluchistan, where his wife now lived. He spent enough time that Latifa gave birth to a daughter, Sana, the following year, but soon reemerged as a man about town in Peshawar and Karachi, a kind of folk hero much sought after by people who wanted to blow things up.

Yousef's brother, Abdul Karim, lived in Karachi and his uncle, Khalid Sheikh Mohammed, was in and out of town from Doha, which is just an hour away by air across the Arabian Sea. Karachi was a tough, broad-backed, sprawling place, with a huge Baluch population (approximately 2 million of its 12 million people) and a large Arab community. Where Peshawar was ancient and connected by centuries of trade and tradition to the Central Asian interior, Karachi was new, created in the nineteenth century and connected to the wider world. It had a deep-water port and an international airport with connections throughout the Gulf, to Europe, Southeast Asia, Africa, and the Americas.

Karachi was a reasonably modern, at times almost normal place, Pakistan's most cosmopolitan city, if also its most violent. Kids on bikes pedaled to school in the mornings. Boys and girls giggled in one another's company at bus stops, and older children listened to music that offended their parents' ears. Men wore midcalf cotton chemises and billowy same-colored pants, most of them grays and whites with only the occasional dandy splashed in violet or a sunburst of bright gold. Many wore a sort of close-fitting hat typical of the Sindh Province, of which Karachi was the capital; most also wore sandals and beards. Women were dressed in similarly styled tunics and pants, but theirs were patterned deep greens and blues and made of slippery shiny silk. In the Baluch neighborhoods the colors were brighter, more dramatic and many. Karachi had some of the vanity and swagger common to cities accustomed to dominating their surroundings. It was, in that way, in love with the myth of itself as a place of danger and deception.

It was an easy place to blend in. The Baluch neighborhoods, in particular, were places you could get lost—if you wanted. They were by far the densest places in a dense city. The streets were paved but covered with an inch-thick layer of dirt and garbage. Horses, donkeys, and goats and horse-drawn cabs slowed the big buses overloaded for the trip home to Quetta or the villages in the Makran Range en route to Iran. The sidewalks were crammed with carts full of crates of thick oranges and red onions and the walkways were messy with refuse from open-air butcher shops.

The city was by many measures a mess. It hadn't had a comprehensive development plan since the 1920s, and the air was foul, though cars were small enough that traffic was manageable. Important people rode Toyota Corollas to work, some with chauffeurs and bodyguards. When Khalid Sheikh arrived Karachi was in the process of earning a reputation as one of the most violent cities in the world. Kidnapping for ransom became so common that authorities were able to stem it, if only slightly, by adopting the desperate strategy of kidnapping kidnap suspects' relatives and then trading hostages. Signs were posted in public places advising "what to do in case of kidnap," as if it were—like an earthquake—a natural occurrence. Murder was equally common. There were well more than a thousand every year. A single month in 1995 had 276 murders, most the result of sectarian rivalries that made parts of the city the exclusive property of one religious political party or another. Most of the killings involved Sunni Muslims killing Shiites or vice versa. The districts each sect controlled were determined to be "no-go areas," not just for political rivals but also for police, who abandoned any pretext of controlling what went on within them. The violence was so normalized that it was common to see the no-go areas referred to in the daily newspapers as NGAs.

Everybody understood what it meant. The NGAs became part of the fabric of the city, a part of the tenor of the time that made Karachi an ideal place from which to marshal a war.

Karachi was the worst but sadly not the only city in Pakistan soaked in blood. The country, conceived as a religious homeland, a place separated for Muslims to escape the majority Hindus of India in 1949, has suffered since its beginning a type of religious strife, or, on many levels, warfare, ever since. There has been a degree of violence here that would be almost unimaginable in the West. The sectarian killings were overlaid on a backdrop of an already violent society, one where honor and shame dictate events that here were normal, elsewhere an outrage.

Here was a typical murder from a typical day's newspaper: A 55-year-old man was shot dead in Samnabad. Police said Gulab Khan, a donkey-cart driver, was intercepted near Al-Noor Mor by a man who came there in a rickshaw. Those living in the area told police that the unidentified man got down from the rickshaw, had arguments with Gulab Khan, shot him, and sped off in the rickshaw. Gulab Khan's family told police that some thirty years back he had killed a man of his rival clan. Police suspected the thirty-year-old vendetta was the motive for the killing.

Here was another: Zulfiqar Ahmad shot to death his sister and her lover in Mandiala Village. Farzana, the sister, had come to her parents after the death of her husband. While there, she developed relations with Hamid of Lahore. On a Monday, they were sitting together when Zulfiqar came and shot them. They both died on the spot.

Or a third: A man axed to death his sister and her alleged paramour in Budhani Village on Sunday. The killer, Mohammed Ramazan, who surrendered to police, said he caught his sister and her lover, Jalil, in a compromising position when he returned home.

And on and on. It was what day-to-day life in Pakistan had become. Khalid Sheikh kept an apartment in the Sharafabad neighborhood, not far from Binori Mosque and school, which was well known within the city for its militancy. (Mullah Omar, future leader of the Afghan Taliban, was a student at the school.) The neighborhood itself is of a kind with others Khalid lived in at different times—downscale commercial, full of cheap hotels, transients, travel agencies, and money changers. It was the kind of place you could come and go unnoticed in the blur of identities. Khalid told people he was Abdul Magid, a Saudi businessman who dealt in a variety of import and export products. Among those, he said, were Japanese electronics and Saudi Arabian holy water.[24]

He traveled extensively, to China, Bosnia, Brazil, Sudan, and Malaysia.[25] Little of this had to do with his work as public works engineer. Most, if not all, of this travel appears to have been related to his abiding interest in carrying out terrorist operations.

Khalid had kept in close contact with Yousef in New York, talking often by telephone. He was impressed by the renown Yousef had gained with the attack and saw it as a model for a future course of action.[26]

Yousef, despite being the object of a worldwide manhunt, was eager to get back to work, brimming with ideas. His old friend from Fahaheel, Abdul Murad, had come to Karachi, trying to get a job with Pakistani International Airlines. In truth, he would have taken a job flying for anyone who would hire him, but after having spent three years attending five flight schools and finally getting his pilot's license, he never found a job as a pilot. Instead, he started talking with his old friend, Yousef.

Murad, as he often did in Karachi, stayed with Yousef's brother, Abdul Karim. Yousef, when he was in town, stayed at the Embassy Hotel. Apparently, he and his brother did not get on that well.

Yousef and Murad met at the Embassy, or they'd go out for meals. They talked about the need for good Muslims to give their lives, if need be, to the struggle,[27] an idea Murad initially balked at, then came to embrace. They talked about potential targets: Benazir Bhutto, then a candidate for prime minister of Pakistan; nuclear power stations; a government official in Iran; the U.S. consulate there in Karachi, and a variety of other U.S. government buildings. There was a plan to assassinate Bill Clinton. Murad proposed packing a small airplane full of explosives and dive-bombing into the Pentagon or the headquarters of the Central Intelligence Agency. Yousef liked the idea and said it was worth considering. He took Murad to meet Khalid Sheikh, introducing Khalid as Abdul Magid, the Saudi businessman. Khalid was then 29, but looked much older and could reasonably play the part of an older man. They met at the Sharafabad apartment and talked at length about Murad's flight training. Khalid wanted to know the details of how long it took, how expensive it was, and who could qualify to do it. They met a second time at a restaurant on Tariq Road, not far away, and again Khalid pumped Murad for information about flying. He told Murad he had a visa for the United States and was thinking of enrolling in a flight school there. Murad recommended a school in Albany, New York, one of several he had attended. When they met a third time, another man, an Afghan named Wali Khan Amin Shah joined them. Shah and Khalid Sheikh knew one another from Peshawar, where one of Shah's good friends was Osama Bin Laden.[28]

Later that year, after Murad had moved into a Karachi apartment owned by an uncle's friend, Yousef came to stay with him. He had a plan to assassinate Bhutto with a remote-control bomb.[29] Khalid Sheikh obtained the funding and the necessary equipment from Sayyaf's camp outside Peshawar. It was delivered to Murad's apartment, where Yousef began to build the bomb, making chocolate, as he described it. In the midst of his preparations, as he was trying to clean lead azide—a volatile substance used to make detonators—out of a container, it exploded in his face. Murad and another friend involved in the plot rushed Yousef to the nearest hospital, but hustled him out when the staff began asking how he had received the wounds. They took him to a second hospital where Yousef told the staff he had had an accident with a butane gas canister. Surgeons were able to save partial sight in his injured eye. As Yousef recuperated, Khalid Sheikh showed up to pay the medical bills.

After he was released from the hospital, Yousef went home to Baluchistan to recuperate. Unwilling to lie low for long, the peripatetic bomber flew to Bangkok, Thailand, and began recruiting a crew to bomb the Israeli embassy there.[30] It was, once again, a slapdash affair involving untrained men and auto accidents, and in this case an abandoned bomb that was discovered unexploded in the back of a stolen truck weeks later. It came to nothing and Yousef returned again to Karachi. He met again with Murad, and this time persuaded his friend to join the cause. The two of them moved to an open-air warehouse compound in Lahore, where Yousef spent almost three weeks teaching Murad to build bombs. This time, it was being done with a purpose. He and Khalid Sheikh had devised a new plan.

Khalid Sheikh Mohammed, a Pakistani born in Kuwait and educated in the United States, had been sought since 1996 by the United States for his role in a foiled airline bomb plot. He was close to capture at least once but even leading investigators had no idea he had joined forces with Osama Bin Laden and planned the September 11 attacks.

Khalid Sheikh Mohammed taught at the University of Dawa al Jihad, which means Convert and Struggle, outside Peshawar, Pakistan, in the late 1980s. The university was founded by Afghan warlord Abdur Rasul Sayyaf and funded in large part by the United States and Saudi Arabia. *(Photo by Abdul Majeed Goraya)*

Many of the 9/11 hijackers were young men who were recruited to the cause of radical Islam by clerics in their hometowns; fourteen of the nineteen were Saudis. Many had left home to fight against the Russians in Chechnya but were selected to join the 9/11 attacks at training camps in Afghanistan. *From top left,* Ahmed al-Ghamdi, Hamza al-Ghamdi, Saeed al-Ghamdi, Nawaf al-Hazmi, Salim al-Hazmi, Khalid al-Mihdhar, Ahmed al-Nami, Abdul Aziz al-Omari, Mohand al-Shehri, Wail al-Shehri, Waleed al-Shehri, Satam al-Suqami, Fayez Banihammad, Hani Hanjour, Ahmad al-Haznawi, and Majed Moqed.

Mohammed Haydar Zammar, a veteran of the Afghan holy war against the Soviet Union, was a prominent figure of the Islamist scene in Germany. He befriended the group of young students led by Mohamed el-Amir Atta and Omar (Ramzi bin al-Shibh) and is thought to have helped recruit them to go to the Al Qaeda camps. *(Photo by Knut Mueller)*

This group photo is from a home video taken at the 1999 wedding of Said Bahaji at the Al Quds Mosque in Hamburg. Omar (wearing a dark vest over a light shirt) is kneeling in the foreground. The wedding occurred just two months before members of the Hamburg group left for Afghanistan.

Al Quds Mosque, on a seamy street in a tough section of Hamburg, was the center of radical Islam in Germany. It was one of the few Arab mosques in a city where most Muslims are Turks. Three of the 9/11 hijackers worshipped there. They were moderate Muslims when they arrived at Al Quds, fervent martyrs when they left. *(Photo by Knut Mueller)*

Mohamed el-Amir Atta was the oldest of the hijackers and the leader of the hijack teams in the United States. He was the son of a middle-class lawyer in Cairo.

El-Amir Atta first came to Germany in 1992 for graduate school. He eventually earned a master's degree in urban planning from the Technical University of Hamburg-Harburg. The photo is from his initial application for German residence. His family in Egypt was only moderately religious, but he became more fervent and radical in Germany.

El-Amir Atta graduated from Egypt's prestigious Cairo University with a degree in architecture in 1991. That spring he and several classmates went on a trip to Luxor, in southern Egypt, for a senior class project. He left Egypt the next year for Germany. *(Photo courtesy of Mohamed Mokhtar el Rafei)*

Ziad Jarrah and Aysel Sengün met and fell in love at a small university in eastern Germany in 1996. Jarrah later moved to Hamburg, where he became part of a group of young fundamentalist Muslims and eventually went to train in Al Qaeda camps in Afghanistan. This photo was taken in Paris in the fall of 2000.

Jarrah, Mohamed el-Amir Atta, and Marwan al-Shehhi moved to Florida in mid-2000 to attend flight school. All three earned private pilot licenses. Amir and Shehhi also received commercial licenses; Jarrah never completed the work.

Marwan al-Shehhi *(center)* was from a conservative, religious family in the United Arab Emirates. He came to Germany on an army scholarship to study engineering. When he moved to Hamburg in 1999 he quickly became a key member of the group that formed around el-Amir and Omar. He's shown here singing at Said Bahaji's 1999 wedding.

Early on September 11, Mohamed el-Amir Atta *(right)* and Abdul Aziz al-Omari passed through gate security at Portland International Jetport in Maine to catch a commuter flight for Boston, where they arrived just in time to board American Airlines 11, the first plane to hit the World Trade Center.

```
G*L11BOS/ON
AA     11 11SEP   BOS B32   745A  767 ON LIST      F9C19Y63
  1  ALSHEHRI    WAIL        P   LAX   2A-F   1 SC LF CLUB ET
  2  ALSHEHRI.   WALEE       P   LAX   2B-F NB SC LF CLUB ET
  3  MORABITO    LAURA       P   LAX   2D-P  D SC LF TKT BS
  4  RETIK       DAVID       X   LAX   2H-F NB SC LF CLUB AB
                                               ET PPR
  5  ROSS        RICHA       P   LAX   2J-F
  6  NEWELL      RENEE       P   LAX   3A-P
  7  BOUCHARD    CAROL       P   LAX   3B-E
  8  FLYZIK      CAROL       X   LAX   3H-F
  9  PUOPOLO     SONIA       P   LAX   3J-F
 10  HACKEL FAR  PAGE        U   LAX   7A-F
 11  ANGELL      DAVID AC2   U   LAX   8A-F
 12  ANGELL      LYNN  AC2   U   LAX   8B-F
 13  ATTA        MOHAM AG2   J   LAX   8D-F
 14  ALOMARI     ABDUL AG2   J   LAX   8G-F
 15  GLAZER      EDMUN       R   LAX   9A-F
 16  LEWIN       DANIE       R   LAX   9B-F
 17  GAY         PETER       R   LAX   9H-F
 18  HAYES       ROBER       R   LAX   9J-F
 19  HENNESSY    EDWAR       R   LAX  10A-F
 20  AL SUQANI   SATAM       J   LAX  10B-F
 21  WAHLSTROM   MARY  RD2   C   LAX  10H-F
 22  CURRIVAN    PATRI       J   LAX  10J-F
 23  CURRYGREEN  ANDRE       R   LAX  11A-F
 24  MELLO       CHRIS       R   LAX  11B-F
 25  BEUG        CAROL AD2   C   LAX  11D-F
 26  ROSENZWEIG  PHILI       R   LAX  11G-F
```

SUNTRUST BANK, GULF COAST
P O BOX 620547
ORLANDO, FL 32862-0547

SUNTRUST

				WITHDRAWALS/DEBITS
DATE	AMOUNT	DESCRIPTION		
06/28	1,164.00	CHECK CARD PURCHASE		
		UNITED AIR	0162	WOODALE
06/29	104.67	CHECK CARD PURCHASE		
		TARGET	00	DELRAY BEA
07/02	62.99	CHECK CARD PURCHASE		
		SUNGLASS HUT #34		BOSTON
07/02	111.76	CHECK CARD PURCHASE		
		ALAMO RENT-A-CAR		LAS VEGAS
07/02	600.00	CHECK CARD PURCHASE		
		SELECT PHOTO		DELRAY BEA
07/03	21.26	CHECK CARD PURCHASE		
		CRAMPS #4116		MIAMI
07/03	31.78	CHECK CARD PURCHASE		
		PAYLESSSHOESSOURCE		HOLLYWOOD
07/09	55.86	CIRRUS ATM CASH WITHDRAWAL		
		C8		ZH FLUGHAFEN
07/09	55.86	CIRRUS ATM CASH WITHDRAWAL		
		C8		ZH FLUGHAFEN
07/09	279.41	CIRRUS ATM CASH WITHDRAWAL		
		CU		ZH FLUGHAFEN
07/09	558.81	CIRRUS ATM CASH WITHDRAWAL		
		CU		ZH FLUGHAFEN

El-Amir Atta and Omari were joined on American 11 by Wail and Waleed al-Shehri, brothers from Saudi Arabia, and Satam al Suqami, also a Saudi. The men had flown practice flights before September 11 to determine where to sit in the aircraft. A pair of hijackers on each of the four hijacked flights purchased first-class seats as close to the cockpit as they could get. These men rushed the cockpit soon after takeoff. The hijack pilot, seated farther back, followed them up front, and the remaining hijackers guarded the cockpit.

The hijackers lived simply in the United States, staying in a series of low-rent apartments and motels and, according to bank and credit records, shopping mainly at Wal-Mart and other discount stores.

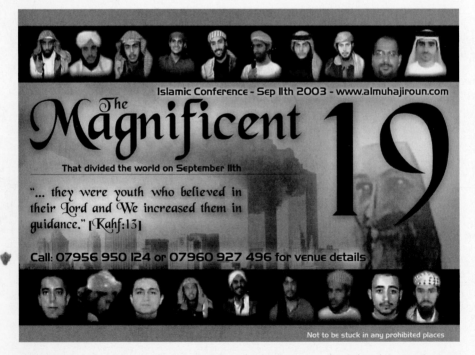

Copies of this handwritten document were found among the effects of several hijackers. It contains instructions on how to prepare for the final suicide mission, including specific directions on how to bathe the night before and how to insure that their knives were sharp.

The 9/11 hijackers became heroes among radical Islamists around the globe. Holy warriors captured or killed on battlefields in Afghanistan carried their portraits. This poster was used to advertise a conference called by the Al-Muhajiroun organization to celebrate the uprising of Islam in London on the second anniversary of the September 11 attacks.

KUALA LUMPUR

Fundamentalist Islam had long been viewed as a greater threat in the Muslim world than it had been in the West, which tended to refrain from criticizing anyone's belief system even when it might be antagonistic to the society hosting it. Many mujahideen came home from the Afghan war to inhospitable regimes. Some of the men were lost, without purpose except to carry out the jihad. One such man, with his wife, arrived at a compound of migrant quarters in tiny Kampung Sungai Manggis, south of Kuala Lumpur, Malaysia, in early April 1992. He was short, stout, heavily bearded, and wearing a skullcap; she was even shorter, and completely covered in a dark burka and full veil. The couple were strangers to Mior Mohamad Yuhana, the man who owned the migrant shacks, but they came in the company of the respected head of the sizable local Indonesian community, and Mior thought they looked kindly.

The visitor, an Indonesian, said he was moving to Malaysia to escape Indonesian President Suharto's repression. He said his name was Hambali. All he wanted, he said, was to be able to practice Islam more freely. Mior told him he didn't care about the man's religion or politics. In fact, two other Indonesians who rented houses in the compound said they were there for the same reason. Stay out of trouble, pay the rent, and we'll be fine, Mior said.[31] He led the couple to a wooden shack, about the size of a one-car garage, with weathered unpainted siding, a bare concrete floor, and a single light bulb inside. They moved in.

Hambali had been born Encep Nurjaman in the highlands of west Java, the central island of the Indonesian archipelago. Indonesia, the fourth most populous nation on Earth, was by far the largest majority Muslim state. The religion was often cross-cut there with animist strains of Buddhism, Hinduism, and Christianity, but, especially on Java, it was the dominant faith. Hambali, quiet and studious, attended an Islamic boarding school (called *pesantrens* there, but very similar to the madrassas of Pakistan) and university. He answered the call to jihad in the late 1980s and spent three years in Afghanistan, where he trained at Abdur Rasul Sayyaf's camps and met, among others, Khalid Sheikh Mohammed. It was also where he took his new name. He returned to Indonesia as Hambali, then, finding his new militancy unwelcome, fled to Malaysia with his wife, a Chinese Muslim he'd met at a religious school where he taught.[32] They arrived in Sungai Manggis with all their worldly possessions—the clothes they wore and a single small bag each.

"They had nothing else when they moved into the vacant house. No sofas, no chairs, no tables—nothing. They cooked and ate, slept on the floor," Mior said.

Sungai Manggis was just minutes from the western Malaysian coast, and from there an hour by smuggler's speedboat across the Strait of Malacca to the Indonesian island of Sumatra. The coastline was dotted with coves and the mouths of small rivers and was a well-traveled path for poor Indonesians, who came for work. A tributary of Kuala Langat River, which runs into the straits, flows directly behind the kampong. It was a rural but hardly bucolic place. Little factories were springing up all over, some so tiny they fit in a backyard and employed just handfuls of people. But Sungai Manggis was not a place to get rich. The area was blanketed with rubber plantations, the economic prize that had brought the British Empire to the region. The plantations had been largely abandoned when the world market moved on to synthetic rubber substitutes and were now overgrown and being reclaimed by jungle. The landscape was green and tangled, the earth a deep orange clay that clung to everything as dust in the morning and mud after the heavy regular midday rains. The rains offered little relief from the oppressive heat. As it fell on the warm ground, a blanket of steam would rise, wrapping everything in a wet, suffocating embrace. The hills were empty as yet of the Western-style subdivisions of the capital, but the bulldozers were coming. The area was being pulled into the sprawling compass of Kuala Lumpur. Roadside stands, piled high with mangoes, pineapple, and durian, also carried an indication of the oncoming march of the suburbs—sacks of used golf balls.

Hambali did odd jobs and showed an entrepreneurial bent. He began appearing outside the gold-domed mosque on the southern edge of the nearby market town of Banting, selling kebabs out of a pedicab cart. His wife, Noralwizah Lee, known around the village as Awi, was joined by her mother, who moved in from their native Borneo. The women seldom ventured far beyond the rented shack. Hambali switched from kebabs to patent medicines, a honey-and-mineral powder mixture he imported that was said to be good for indigestion. He prospered. Soon, he was driving a blue Proton hatchback and juggling calls on a pair of cell phones. Some of the calls were made to an Arab man who had recently arrived in Manila, Osama Bin Laden's brother-in-law, Mohammed Jamal Khalifa.[33] Khalifa came to Manila in 1988 and began establishing a network of small businesses—a rattan furniture factory, a travel agency, and others—as well as a network of Islamic charities. By 1992, Filipino intelligence officers had begun to suspect he was using the charities, at least in part, as fronts to direct money to Islamist rebels in the southern province of Mindanao.[34] Informers from within the separatist Abu Sayyaf group told authorities Khalifa was one of their key financiers. Telephone taps revealed Khalifa's frequent conversations with Abu Sayyaf officers as well as Hambali, who shared the Sayyaf goal of a pan-national Muslim state.

Hambali applied for and received a contractor's license and employed small work crews to do small construction jobs in the area. He began traveling—on business, he told his landlord—disappearing for weeks at a time. At home, he received what became a steady stream of visitors. They spoke English and Arabic and sometimes carried duty-free shopping bags. The men were "in their late twenties or early thirties. They looked tough. I remembered thinking at that time they would make good footballers," Mior said. The men, as it would turn out, were jihadis coming or going to the Afghan training camps, sent there by a militant organization founded by Hambali and two other men who lived nearby— Abu Bakar Bashir and Mohamed Iqbal. The organization was called Jemaah Islamiyah. JI, as it came to be called, over time sent hundreds of men to the camps.[35] When they came home many were integrated into an organization that sought to convert all of Muslim Southeast Asia into an Islamic state.

"He was so humble, very clean and open face. You don't have a second thought about trusting him. Not an Islamic militant at all," Mior said of Hambali. "Sometimes Bashir and Iqbal joined in the prayers at the prayer room [in the village]. Iqbal usually leads the prayers, sometimes Bashir. Hambali never leads the prayers. Iqbal and Bashir used the Muslim-Christian clashes in Maluku and Ambon in their lectures in the [prayer room] as illustrations of Muslim oppression. They mention jihad but did not ask us to take up arms."

Mior was wrong, of course. As unlikely as it seemed, Sungai Manggis—a tiny village in the wet red-clay middle of nowhere—had become the center of Islamist militancy in Southeast Asia.

MANILA

Hambali had remained in contact with men from his own jihad days, among them Khalid Sheikh Mohammed. In the spring of 1994, Mohammed sent Wali Shah to meet Hambali in Kuala Lumpur. Shah and Hambali incorporated a trading company called Konsojaya, saying they intended to go into the palm oil business, exporting it to the Middle East. Its real business was as a conduit for money Mohammed would raise in the Middle East and send out to East Asia. Ramzi Yousef and Shah had been in and out of Hambali's region several times, working with Abu Sayyaf in the southern Philippines and visiting Manila often enough to acquire steady girlfriends. In Yousef's case, the girls included a pretty young dancer at a local go-go club and a sales clerk at a Kentucky Fried Chicken outlet.[36] Shah developed a long-term relationship with a bar girl, Carol Santiago. Mohammed came in and out of the region, too. He never had a steady girlfriend, as such, but asked young women he met to do favors for him. These included opening local bank and cell phone accounts. He paid them for their time.[37]

Mohammed, Yousef, and Shah returned to the Philippines in early 1994. Once in the country, they split up. Shah moved in with Santiago at an apartment on Singalong Street. Yousef, using the name Adam Baluch, rented a room at the third-class Manor Hotel. Mohammed, under the name Salem Ali, rented an apartment in the Tiffany Mansions, a new, thirty-five-story condominium building in the Greenhill section of the city. The Tiffany had an airy marble lobby, a gym, a pool, and a Jacuzzi spa. One of Mohammed's neighbors was Joseph Estrada, a Filipino film star who was soon to become president of the country. Mohammed bought a Toyota Corona sedan and left in it every morning, dressed in polo shirts and casual slacks, as if going to the office.

They moved in and out of town. They met in public places—convenience stores, hotels, karaoke clubs, and go-go bars—where they would be inconspicuous. Police had wiretaps on dozens of suspected Islamist terrorists in the city, but they never caught a trace of the three men. Only afterward, searching through the trash outside Shah's apartment, did they discover that the shredded lease was made through Bin Laden's brother-in-law, Mohammed Khalifa.[38]

On an earlier trip to Mindanao, at the southern tip of the island nation, Yousef and Abu Sayyaf leader Abdurajak Janjalani had concocted a seemingly far-fetched plan to assassinate Pope John Paul II, who had announced plans to visit the Philippines in early 1995.[39] Yousef confided to Murad that he thought the Abu Sayyaf men were too unsophisticated to do anything even approaching killing a pope, but he thought it was such an excellent idea he would use Janjalani's resources to do it himself, then give Abu Sayyaf credit.[40] The plot against the pope wasn't even the half of it, however. Yousef also wanted to assassinate U.S. President Bill Clinton, who was due in Manila for a brief visit before the pope. He abandoned the Clinton attempt only after researching the preparations for the visit and concluding that security for the American president would be too tight and the opportunities too few.

The pope seemed more vulnerable. For one thing, he intended to stay in the country for almost a week and would be speaking to several large public gatherings. Papal masses frequently drew crowds in the hundreds of thousands and enough tumult to mask almost any activity Yousef and Mohammed might undertake. They considered several options: a simple shooting, an attack from the air, and a suicide bomb attack at one of his outdoor events. Yousef and Mohammed at one point flew around Manila in a rented helicopter. Yousef told Murad this was just Mohammed trying to impress a woman he was courting, but it seems more likely it was an attempt to surveil the area. The air attack was discarded when Yousef discovered the Philippine government intended to restrict airspace in the vicinity of the pope's events for the full length of his visit. They bought priests' cassocks and Bibles to disguise themselves in an attempt to get close enough to explode a bomb in the pope's presence, but this, too, was discarded. They looked at the possible routes the pope would be traveling and in the end settled on an attack using remote-controlled pipe bombs, which could be exploded at a distance. In November, Shah and his girl-friend moved to a room at the Dona Josefa apartments, which sat along-side the likely route to and from the papal ambassador's home where the pope would be lodged while in Manila.

The Josefa sat on the seam of two worlds—the leafy residential precincts of gracious and quiet older homes where the papal nunciature was located and the old city and government center. In between, was the Josefa's Ermita neighborhood, a brightly lit, all-hours entertainment dis-trict filled with huge billiards clubs, three-story karaoke parlors, and bars. With its money changers, international telephone centers, barber shops, beauty salons, a halal butcher, and airline-ticket shops, it bore consider-able resemblance to Mohammed's Karachi neighborhood and even the area of Fahaheel, Kuwait, where he grew up. There was a fluidity in these places, an anonymous ease with which people of different cultures and languages mix and pass by on their way, always, to someplace else. President Clinton liked to say that international terrorism was the dark side of globalization. In these neighborhoods, the evidence underlying that idea was on display daily. And nightly, because Ermita more than anything was a nighttime neighborhood.[41]

Shah's fourth-floor studio apartment at the Josefa cost less than $300 a month. For this he received a 30-foot-by-12-foot room with parquet floors, built-in kitchen appliances, and a bed. The lobby had terrazzo floors, plaster walls, and the world's tiniest atrium. The rare breezes were stirred mainly by whatever air the trucks and taxis ruffled along the six-lane road out front. The Josefa's chief recommendation to the plotters was that road, which was in the middle of the pope's parade route.

Yousef had been experimenting with different explosive devices. He built the crudest sort of pipe bombs, nothing more than a length of metal tube packed with explosive. Eighth-grade boys could make them. At the other extreme, he was on the verge of perfecting a small bomb he believed could be disassembled into such tiny, innocuous components that they could elude the most stringent security. It consisted of an amount of nitro-glycerine small enough to be stored in contact lens solution bottles or toothpaste tubes. The detonator that would ignite the nitro would be connected to a timer built out of a modified Casio Databank watch. The beauty of the Databank, for the bomber, was both its ready availability and the ability of its alarm to be set for up to a year into the future. Yousef modified the watch, installing a tiny electronic plug that, unless someone knew the watch intimately, would not be detected by casual observation. It looked like a normal watch. A small fuse made from two 9-volt batteries could be attached to the plug and to the bomb in minutes. The disassembled bomb works could be carried through airport security magnetometers without setting off alarms or suspicion. Then it could be assembled and set in minutes in the airplane lavatory. If set for hours or even days ahead, the bomber could be in the next county or continent by the time it went off.

By December 1, Yousef had a working prototype and was ready to test it. He gave the preassembled bomb, easily carried in a knapsack or shopping bag, to Wali Shah with instructions to take it to a movie theater near Mohammed's condo and attach it to the bottom of a theater seat, then leave. Shah did as instructed. A short while later, the tiny bomb exploded. The charge was small enough no one was injured, but the demonstration was successful. Everything worked exactly as anticipated. Now they were nearly ready to act.

Yousef and Mohammed had come to the Philippines because of its fledgling support network of Islamists and because it was cheap, a place they considered as a potential base to develop Yousef's still imaginary Liberation Army. But they had no desire to attack Filipinos. If anything, they might have sympathized with the locals. Since the oil boom, Gulf states, like Kuwait, where Yousef and Mohammed grew up, have been filled with foreign workers from South Asia—Pakistanis and Indians— and Southeast Asia, mainly Filipinos. There were more Filipinos in Kuwait, the UAE, and Saudi Arabia than there were Arabs. Filipinos were the service class, the shopkeepers, maids, and sales clerks. If anything, the Filipinos occupied an even lower rung on the social ladder than did Kuwaiti-born Pakistanis like Yousef and Mohammed.

What Mohammed and Yousef wanted to do was attack Americans, but with the Clinton plan foregone, they needed new targets. The ideal target for Yousef's tiny new bombs would be one that maximized the effect of a small explosion. Except for the intended result—death—this would be the opposite of putting a huge fertilizer bomb in a truck and driving the truck into the basement of a big building, as Yousef had done in New York.

They found the perfect objective—commercial airliners. On trans-Pacific routes, airlines, to maximize profits, typically flew the biggest airplanes they could get their hands on. In those days, that gave the airlines basically one choice—Boeing's hugely successful 747 jumbo jet. Throughout East Asia in the 1990s, every international airport looked like a Boeing sales lot. Parking aprons, taxiways, the air were all filled with the big jets. They hopped from country to country, filling up for long hauls across the Pacific. In some countries they were even used for short-haul flights. The planes were big, they contained a lot of people, they were vulnerable. Nowhere could small bombs have more effect.

Mohammed and Yousef had conceived the idea the previous summer and patiently put together the team to execute it.[42] Using nothing more exotic or complicated than airline timetables, they devised a scheme whereby five men could in a single day board twelve flights—two each for three of the men, three each for the other two—assemble and deposit their bombs, exit the planes with the timers set to ignite the bombs up to several days ahead, allowing the men to be far away and far from reasonable suspicion by the time they exploded. The math was simple: twelve flights with at least 400 people per flight—somewhere in the neighborhood of 5,000 deaths. It would be a day of glory for them, calamity for the Americans they supposed would fill the aircraft.

Immediately after the cinema test, Yousef called his friend Murad, now in Dubai, and said he had a job for him.[43] He wanted him to come to Manila as quickly as he could. He would wire money for the airfare. Murad agreed to come. He didn't even bother to ask what the job was. He knew. Yousef hadn't bothered to give him the private three-week bomb-making class without a reason.

Yousef and Mohammed needed one further test to ensure they could get the bombs on board an airplane. Yousef took this practice run himself. On December 8, he flew Philippine Air Flight 434, scheduled from Manila to Tokyo with an intermediate stop in Cebu. Once on board, Yousef went to the men's room, assembled his little bomb, returned to his row, attached it beneath the seat, then exited the plane at Cebu. An hour later, the bomb blew up in midflight, killing passenger Haruki Ikegami. It was only through great skill that the pilot kept the whole plane from careening into the ocean. Yousef knew he could make a slightly bigger bomb that, well-placed, an airplane would never survive. And now he knew he could get it in place.

The two practice bombs contributed to a time of already extraordinary anxiety in the Philippines. The nation—never far from implosion—seemed on the verge of losing a long battle in its southern provinces with the increasingly violent Islamic separatist movement. Insurgencies of one kind or another had been ongoing in the south for decades, but this most recent version was more deadly and persistent. The national police had just completed a 182-page catalog of the previous twelve months of terrorist activity. By any measure, it had been a horrible year: more than fifty attacks resulting in 101 deaths, with Roman Catholic priests among the frequent targets.[44]

There were fears within the local intelligence services that the separatists might use the papal visit as a stage to call the world's attention to their cause.[45] Some analysts thought they might have more in mind than a simple demonstration, perhaps going so far as an attack on the pope. They instituted Operation Santo Papa to investigate.[46] When a typhoon barreled through in mid-December, ripping out palm trees, houses, and power lines, it seemed almost inconsequential.

There was no sign the nation at large was taking much notice. It was, after all, Christmastime. Roman Catholic holidays are serious undertakings in the Philippines, which for 300 years was a Spanish colony and remains Asia's most Catholic country. In normal years, Christmas is treated as a period of prolonged and unavoidable celebration, an interlude in what can be the long hard grind of everyday people's everyday lives. Gift buying begins in summer; colored lights and Christmas lanterns get hung in September, about the time radio stations start playing Christmas carols. Midnight mass—in much of the world an event reserved for Christmas Eve—is celebrated nightly for weeks on end. In a country defined largely by its poverty, the season is an extravagance, recognized and embraced as such.

The time was right for the bombers. Shah and Santiago moved out of the Josefa. Yousef moved in, taking an identical room on the top floor, facing directly onto President Quirino Avenue. He began gathering the supplies he would need to build his armory. He had false papers identifying himself as an English chemist and used them as he made the rounds of chemical supply houses. As weeks went by, Yousef became so secretive he stopped letting the room boy in to change the sheets. He and Shah hauled enough boxes and bottles through the Josefa lobby that the staff started to wonder what was up. Manila, on its kindest days, was a steam bath, a place that drenched you with humidity in the first five minutes and drowned you in the next. With oppressive heat and a ready supply of cheap labor, visitors, if they could avoid and afford it, didn't exert themselves much. So when the tenants in room 603, who the staff took to be young Arabs—a category of visitor even less inclined to physical labor than others— hauled those boxes and bottles through the Josefa lobby, they generated a level of interest that in simpler times, they might not have. The Josefa was not usually the sort of place where people scrutinized strangers. The city had for decades been a regular stop on the sexual tourism circuit ridden by wealthier East Asians, Americans, and oil-rich Middle Easterners, and the Josefa was at the southern end of Ermita, one of the city's evening entertainment districts. In normal times, its cheap, clean rooms were filled with men of modest means and desires.

Not only were these Arabs doing physical labor, rare enough, but once you started looking, it was hard not to notice how banged up they were. Khan was missing three fingers on one hand—a result of the Afghan wars; Yousef's hands and face were scarred and he had that odd eye. (Little did the staff know: Shah's entire body was scarred and Yousef had burn marks on his feet and his back.)

With all the nervousness about the pope's imminent arrival, it wasn't long before someone passed word to the local ward heeler, Apolinario Medenilla, that something odd seemed to be going on at the Josefa. Medenilla, in turn, gave the local police a heads-up about the suspicious men, but there was little the police could do about it. "There was enough to be suspicious about, not enough to act on," said Manila Mayor Jose L. Atienza Jr.[47]

The plotters went about their preparations with little idea anyone had noticed anything. On December 21, they gathered friends together for a party celebrating the anniversary of Pan Am 103, which had been blown up by Libyan agents over Lockerbie, Scotland, killing 259 passengers just six years before. It was almost certainly the only celebration of the event in Manila that night.[48]

Murad flew in the day after Christmas. As instructed by Yousef, he booked a room at a nearby hotel, Los Palmas. He checked in under a name Yousef had given him, ignoring Yousef, who was sitting in the lobby waiting. He went up to his room and Yousef followed minutes later. Murad, as requested, had brought along a half-dozen different shades of L'Oréal hair dye. Yousef wanted to change their appearances. Murad noticed that Yousef had already started changing his. For the first time in years, he was clean-shaven. Within minutes, so was Murad. They stayed at Las Palmas for two days, then Murad checked out and moved in with Yousef at the Josefa. Murad was not as surprised to see all the bomb-making materials as he was to see the man he had been introduced to as Abdul Magid the year before in Karachi, Khalid Sheikh Mohammed. Yousef and Murad went to work buying the last of their supplies and began on the bombs. They were still simultaneously pursuing the plot against the pope, so they bought, in addition, priests' cassocks, crucifixes, and Bibles to use as disguises. Whenever Mohammed visited the apartment he wore gloves. He said he was wary of the chemicals. He might also have been wary of leaving fingerprints, which he very carefully did not; no one but Yousef knew his true name, and he seemed to have a different identity for every other person he met. He wasn't taking any chances.

As the end of the year approached, Yousef and Mohammed left Murad at the Josefa with instructions to continue work on building the bombs and—or so they told him—went to take scuba-diving classes at a nearby coastal resort. Yousef returned alone on January 1. Murad never saw Mohammed again. He and Yousef spent the next several days assembling timers and detonators. Yousef told Murad to plan on flying out to Singapore to start his bombing run from there on January 14. Yousef never suggested there were plans for four other men to put explosives aboard ten other airplanes.[49]

The two of them developed a routine: they slept in, often past noon;

ate lunch; shopped at the mall; maybe walked along the waterfront. In the evening they would watch CNN, talk about the poor condition of the Islamic nation, and build bombs. January 6 was more or less the same until evening. Yousef was disposing of extra chemicals, things they wouldn't need, by burning them off in a pan on the stove top. Something went wrong and huge clouds of smoke engulfed the room. They opened the windows, but it wasn't sufficient. They escaped into the hall. The neighbors by now had noticed and asked what had happened. Yousef tried to explain that they were making firecrackers for a belated New Year celebration. Firemen came, took a look around, and left. A cop came by, Yousef drifted away, and Murad answered the cop's few questions, then he left, too.

Things seemed to be settling down when the fire alarm, which until then had been silent, sounded.[50] This time police and firefighters rushed to the scene. They let themselves into room 603 and found a scorched pot on the stove top and the place littered with beakers, funnels, cotton batting, cans of gasoline, and, in the refrigerator, a pair of king-size Welch's grape juice bottles filled with what turned out to be liquid nitroglycerin.[51] The police locked the place back up and went hunting for a search warrant. Not wanting to reveal the unauthorized entry, they had a tough time justifying a search of the premises. Eleven different judges denied their request, but the twelfth signed it and investigators returned to the Josefa.

They spent the rest of the night and much of the next day sorting through enough evidence to eventually fill three police vans. Before the search ended, two generals, a colonel, a major, a captain, a lieutenant, and a dozen other police and army officers had joined the crowd. Their haul included priests' cassocks and collars, Bibles, crucifixes, two cartons of Roughrider lubricated condoms and a map of the pope's prospective travels; chemistry textbooks and chemicals—acids and nitrates by the gallon, one finished pipe bomb and another waiting to be packed; soldering irons, switches and loops of electrical wire; a dozen passports and as many Casio Databank watches.[52]

Murad, after he finished with the earlier police questions, had called Yousef's cell phone and been told to get to the 7-Eleven on Mabini Street. When Murad arrived, Yousef was standing there with another man from

Karachi, Wali Shah, whom Murad knew as Osama Asmurai and whom he hadn't seen since the summer of 1993. They went to a karaoke bar to decide what to do. Yousef was most concerned about his laptop, and Murad agreed to go back and retrieve it. When he returned, he saw the police still hanging around the Josefa lobby. He paused. One of the security guards saw him, too, and gave a shout. Murad spun on his heel and attempted to flee, tripping and falling and getting caught. That was the end of it. Shah was caught the next day coming out of his girlfriend's apartment. Yousef, meanwhile, walked off into the night, and the next day flew to Thailand and from there on to Pakistan. His participation was not long a secret, however. Murad, in a vain attempt to save himself, practically begged his interrogators to listen to what he knew about the World Trade Center. "We hit a gold mine with the arrest of Murad," said Senior Superintendant Rodolfo "Boogie" Mendoza. "We threatened him that we were going to turn him over to the Mossad. I told him he had nothing to tell me, but maybe they would be interested.

"'You're a shit, a nothing to me,'" Mendoza told him. "He insisted he was important. 'Please listen to me. I was involved in the World Trade Center and I know Ramzi Yousef.'"[53]

This revelation, which audiotapes later revealed was made under considerably more duress than Mendoza mentioned in his account—he had put a hose in Murad's mouth and force-fed water down his throat, effectively starting to drown him as he sat on a chair in an interview room—didn't save Murad. After a few weeks, when he had gotten everything he thought he could get, including Murad's prideful admission that he had proposed hijacking an airplane and dive-bombing it into the CIA headquarters, Mendoza handed Murad and the information over to the Americans. A worldwide manhunt for Yousef ensued. Within months, Yousef's whereabouts were betrayed by another man he had tried to recruit into yet another new plot. He was nothing if not persistent. He was captured in a raid by Pakistani security forces as U.S. agents watched at a small hotel in Islamabad, the Pakistani capital. Shah, who had bribed his way out of a Philippine prison just three days after his initial capture, was arrested again seven months later on a resort island in Malaysia where Hambali had arranged for him to hide out. That left just one of the known

Manila plotters at large—Khalid Sheikh Mohammed. Investigators chased leads all over the globe but had no real fix on where or even who he was. At that point, they knew so little about him, they didn't even know it when they came close to catching him. How much did they not know?

When, in February 1995, American and Pakistani agents stormed the hotel in Islamabad and hauled Yousef away, kicking and screaming, it was a frightful scene, according to another guest at the hotel.

"It was like a hurricane, a big panic," the guest told journalists. "He was shouting: 'Why are you taking me? I am innocent! Show me papers if you are going to arrest me! Who are you?' No one listened to him. They took him without his shoes. His eyes were blindfolded, his head was covered, his arms and legs were tied."[54]

The man giving this account was registered at the front desk as a Karachi businessman. He gave his name as Khalid Sheikh. It was, American authorities eventually came to believe, Mohammed, hiding in plain sight.

CHAPTER 4

War, After War

KHARTOUM

WITHIN WEEKS AFTER THE LAST SOVIET troops trucked across the Amu Darya's Friendship Bridge and out of Afghanistan, Osama Bin Laden was thinking about where he should go as well.[1] The future of Afghanistan was still in play, but Bin Laden was already looking beyond the Hindu Kush. The war the Soviets left behind did not end, or even slow, without them. To a degree, Bin Laden mimicked the Soviets. He continued to fund and help supply the mujahideen—as the Soviets continued to fund and supply the Kabul government—for years, but by mid-1989, he and a small group of other Arabs in Peshawar were planning how they might advance their battle to the next stage. They began to assemble an organization that would provide a de facto answer to the unresolved debate with Abdullah Azzam—whether to carry the jihad outside Afghanistan or stay and ensure complete victory and establishment of an Islamic republic. They resolved to take their fight out into the wider world, and they began searching for a base from which to stage it.

Bin Laden left Peshawar for home in 1989. After Azzam was killed that fall, he took control of the Office of Services and, over the next two years, made periodic, brief trips back to Pakistan. But his attentions had turned elsewhere. He and fellow Afghan-Soviet war veterans—mainly from Egypt and Iraq—formed the core of a small group they called Al Qaeda, and they were busily recruiting others to join.[2] The Egyptians,

who outnumbered the others, made for odd bedfellows with the pious Bin
Laden. More than most who had come to take part in the war, they had
arrived with a developed ideology. It wasn't exactly the same as the one
they left with, but they had a program worked out over a century that
called for all states in the Muslim world to be returned to Muslim doctrine.
The core texts of the standard modern Islamist creed were written by
Egyptians Hasan al-Banna and Sayyid Qutb, one the founder and the
other the great popularizer of the Muslim Brotherhood. They drew on
older teachings but remade them for the modern world. Their main inno-
vation was the insistence on violent revolt against insufficiently Islamist
regimes in the Middle East. The leading Egyptians in Peshawar were veterans of what had become a long campaign against the Egyptian government in which both sides made ample use of that violence. Led by Ayman
al-Zawahiri, Peshawar Egyptians saw Afghanistan not as an end in itself
but as a sort of demonstration project, an experiment to see if a ragtag
Muslim army could defeat the most sophisticated of military opponents.
When the Soviets left, they were ready to declare the demonstration a success and take its lessons forward.

   With Bin Laden as the declared emir, or leader, of the new group,
men were dispatched to inspect countries that might serve as a base of
operations. Afghanistan, after all, was primitive, landlocked, and inhospitable. They needed a place they could operate with safety and with
somewhat more infrastructure than Afghanistan offered. Soon after his
return to Saudi Arabia in 1989, Bin Laden had gathered a force of men
and attempted unsuccessfully to insert them into South Yemen, a Soviet
client state, with hopes of overthrowing another Soviet-sponsored government.[3] The Yemenis uncovered the plot, halted it, and objected force-
fully to the Saudi government, which had its own complaints about Bin
Laden stemming from his increasingly harsh criticism of it.

   One of the areas Al Qaeda had sent people to examine was Sudan,
the largest country in Africa, located directly across the Red Sea from
Saudi Arabia. You could almost see it from Bin Laden's home town of
Jeddah. More attractive still, Sudan's ruling party was one of the few sitting governments in the world that actively supported Islamist causes. The
country was desperately poor and begging for the sort of contribution an

unencumbered investor like Bin Laden could make. By the end of 1990, Bin Laden began buying land in Khartoum, the capital. Over the next two years, he bought farms, salt flats, houses, office buildings, desert parcels, orchards—almost every sort of property the country had to offer. When he was expelled from Saudi Arabia a year later, he already had a place to go—Khartoum. The Al Qaeda operations he had ordered established there became in fact as well as in name his base of operations.

Sudan was so poor that Bin Laden was able, without much apparent effort, to acquire near monopolies on many of the country's principal trading activities. He controlled the gum, corn, sunflower, and sesame trade; invested $50 million in a bank; established by far the largest construction company in the country; built roads, airports, harbors, and farms.[4] He also established businesses that dealt in wheat, palm sugar, salt, soap, furniture, palm oil, peanuts, fava beans, and foreign exchange. He imported heavy-construction equipment and built training camps in the mountain country north of Khartoum. Then he brought in Afghan veterans to run the camps, seeking out jihadis who had difficulties in their home countries. In 1993, he brought in as many as 480 Afghan Arabs who were being expelled from Pakistan.[5]

In Washington, D.C., in discussions of terrorism at the time, Bin Laden's name was mentioned in passing, if at all. The primary terrorist threats were seen as coming from state-sponsored entities, like Hezbollah, the Iranian-backed group that killed 273 U.S. Marines in Beirut in 1983, or directly from states themselves, as illustrated by the Libyan destruction of Pan Am 103. When Bin Laden was talked about at all it was generally as a financial backer of others who were potential dangers. The United States suspected he had funded an attack against an American housing complex in Yemen in 1992, and that he was a consistent source of money for Egyptian terrorist groups and, in fact, had underwritten an attempted assassination of President Hosni Mubarak.[6] It was not known to American intelligence that he and the Egyptians had entered a formal agreement resulting in the formation of Al Qaeda.

"In 1995, '96, '97, Al Qaeda was not on our screens as Al Qaeda. It was one of many organizations we were concerned with," said an American diplomat then assigned to the region. "Osama Bin Laden was looked upon as a sort of venture capitalist of terror. Bring him an idea, he'd give some money, maybe some technical assistance. He wasn't seen as a planner. Beginning in '94, his name kept popping up, but always in the financial sense."[7]

For his part, Bin Laden was growing concerned with the United States. In 1993, although it was not immediately apparent to the United States, he had declared virtual war against American armed forces, which were then in nearby Somalia on a humanitarian mission: "[The] American army now they come to the Horn of Africa, and we have to stop the head of the snake. . . . We have to stop what they do in the Horn of Africa."[8]

By the mid-1990s, Bin Laden had made enough of an impression that the CIA began to track his movements and financial accounts.[9] The United States pressed the Sudanese government to curtail Bin Laden's activities and, when that produced little evident action, to expel him to a country that would allow him less freedom. The problem was finding such a country. The Saudis were approached and wanted nothing to do with Bin Laden.[10] In fact, every country the Sudanese asked, declined. The Sudanese said this included the United States. They claim they offered to turn Bin Laden over to the United States, provided their involvement was not revealed. The United States denies the offer was ever made.[11] There was, in any event, no legal basis since Bin Laden was not then charged with any crime against American citizens.

In late 1995, a National Guard office in Riyadh, Saudi Arabia, was bombed, killing five Americans and two Indians. Again, Bin Laden was suspected of being a possible backer of the attack, but the Saudis announced they had found the perpetrators, and promptly beheaded them all before American investigators could even get into the country to interrogate them.

By February 1996, the Clinton administration was so frustrated with Sudan generally, and its protection of Bin Laden specifically, the U.S. embassy in Khartoum was closed and the staff called home. This sort of action is financially disastrous for a host country; it signals a profound lack of trust in the local ability to protect foreigners, and financial investment usually follows the diplomats out of the country. Then, the United Nations imposed its own economic sanctions, citing the Sudanese connections to the attempted assassination of Mubarak. The Saudis all the while had been urging the Sudanese to do something about Bin Laden. The Saudis were much more influential in the region—and much more willing to reward their partners—than were the Americans. Sudan, finally, realized it had to get rid of Bin Laden and told him so. Bin Laden, hardly pleased, had no choice but to go. He might well have created a stateless organization, but it—and he—physically had to be *someplace*. While in Sudan, he had continued to consult with old allies in Pakistan and inside Afghanistan. He provided the money that kept the Office of Services in Peshawar in business and had remained in contact with some of the various groups running training camps across the border. He made inquiries among them and was told he would be welcome to return.

JALALABAD

On May 18, in an Ariana Afghan airliner leased by the Sudanese government, Bin Laden landed in Jalalabad with three wives, several children, and scores of aides and assistants. He arrived back in Afghanistan just as a new group was consolidating its hold over much of the country. The Taliban, a previously obscure and mysterious army of religious students who championed an absolutely strict interpretation of the Qur'an, had recently arrived themselves as a powerful fighting and ideological force in the country's long internecine warfare. The ascetic leaders of the group promised to cleanse the country not just of its sins but of its past, reasserting the rule of Allah. In some ways, they won control of the country less by virtue of their ideology, and less still by their skills in battle, than by the fact of having fewer enemies and more respect than other warlords, who had been fighting, crossing, and double-crossing one another, and the people, for years. In short, they offered the simplest promise possible—to be an effective police force, stop the fighting, and keep people safe from one another. When the Taliban marched into Kabul the following spring, they took the capital without a fight. Mullah Omar, the one-eyed cleric who led the Taliban, took upon himself the title Commander of the Faithful, and the country was renamed the Islamic Emirate of Afghanistan.[12]

The area Bin Laden came to in Afghanistan was not yet under Taliban control when he arrived. Commanders of mujahideen forces he knew from the Soviet war still held Jalalabad and its hinterland, but the Taliban took it soon after and Bin Laden brokered an agreement with them. In return for their hospitality, he would furnish them experienced fighters and new money. The mujahideen training camps were rejuvenated by Bin Laden's return and the Taliban victories. The Arabs had not had great effect on the battlefield against the Soviets, but in the new, factionalized Afghanistan, Bin Laden's men became a significant military force and also a factory, manufacturing warriors for worldwide adventure.

While the Taliban were consolidating their hold on the country, Bin Laden spent the summer writing the first of what would be several religious decrees, or fatwas. Fatwas were typically issued by religious scholars in an attempt to reconcile Qur'anic principles with some newly arisen problem. Bin Laden was not such a scholar—his education and experience were in civil engineering—but he seemed to be convinced that he had the religious calling, if not the training, necessary to become an authority on Allah's wishes.

This fatwa, like those that would follow, was (at more than 11,000 words) grandiose in conception and tone. Titled "Declaration of War Against the Americans Occupying the Land of the Two Holy Places," it attempted to build a case that the American armed forces were committing a horrible transgression by basing troops on the Arabian Peninsula and should be forcibly expelled. It read, in part:

> It should not be hidden from you that the people of Islam had suffered from aggression, iniquity and injustice imposed on them by the Zionist-Crusaders alliance and their collaborators; to the extent that the Muslims' blood became the cheapest and their wealth as loot in the hands of the enemies. . . . The latest and the greatest of these aggressions, incurred by the Muslims since the death of the Prophet (Allah's blessing and salutations on Him) is the occupation of the land of the two Holy Places— the foundation of the house of Islam, the place of the revela-

tion, the source of the message and the place of the noble Ka'ba, the Kiblah of all Muslims—by the armies of the American Crusaders and their allies.[13]

Bin Laden cited many other injustices inflicted on the Muslim people by the Americans and others of their ilk. He made it clear that Muslims were justified—indeed, commanded—to use any means necessary to expel the Crusaders. It was what Zawahiri, his closest ally, often called "the neglected duty." The Arab-language press, which had helped build the myth of Bin Laden as a warrior prince in the Soviet war, again eagerly publicized his new direction. Videotapes and cassettes of Bin Laden speeches began circulating throughout the Muslim world. As they had done for scores of others, these newer media spread Bin Laden's reputation without attracting great notice in the wider world.

The following June, a fuel truck exploded at a U.S. Marine barracks in Dhahran, Saudi Arabia, killing nineteen people. As before, the Saudis did not afford much opportunity to share in the investigation; they blamed a Saudi offshoot of Hezbollah, rounded up a bunch of suspects and executed them, again without granting the United States an opportunity to interrogate them. Bin Laden did not claim credit for the attacks—but he conspicuously praised the attackers. It became clear, finally, to everyone that simply moving Bin Laden from place to place wasn't going to cure the problem he presented.

DOHA

After his capture, Ramzi Yousef never gave up any valuable information in interrogation. In fact, he was never really interrogated in the way we understand it now. He was interviewed. He gladly acknowledged much of what he had done, even bragged about some of it. And he complained, in effect, about what a tough job he had, how hard it was to put together these plots, to raise the money, to work with these dolts. (When, for example, he was asked why Mohammed Salameh made the fatal mistake of trying to get the deposit back on the Ryder rental truck, he responded with a single word: "Stupid.") What he did not do was tell anybody much that they couldn't or hadn't already figured out for themselves. He did not, specifically, name names. But he had left clues, the laptop computer and his personal telephone directory in particular.

One of the entries in Yousef's telephone directory was for Zahed Sheikh Mohammed, Khalid's brother in Peshawar. Not long after the Americans found the name, Pakistani investigators raided Zahed's offices and those of other Arab charities in the city, but Zahed had fled before they arrived. The raids occurred amid one of the periodic crackdowns on the Arabs by the Bhutto government and added to the huge commotion brewing in the resident population of Afghan Arabs. Many Arabs followed Zahed Sheikh out of town.

Khalid Sheikh Mohammed had been exceptionally cautious on what was his first experience on an actual terrorist operation in Manila. He used three aliases, always wore gloves inside the bomb factory in Room 603, and lived apart. Not even Murad, one of the key conspirators, knew his true identity. There was a reason Yousef had risked sending Murad back for the laptop: Yousef had named names, or at least left them behind. Among the files on the computer was one apparently chastising someone for failing to deliver support:

> To: Brother Mohammad Alsiddiqi:
> We are facing a lot of problems because of you. Fear Allah, Mr. Siddiqi, there is a day of judgment. You will be asked, if you are very busy with something more important, don't give promises to other people. See you on the day of judgment. Still waiting.
> Khalid Sheikh, and Bojinka.[14]

"We knew there was another person involved . . . but he was very mysterious and we didn't know who he was," said Neil Herman, who led the FBI investigation of the Manila plot. "He basically eluded us."[15]

Investigators eventually made the connection between this Khalid Sheikh and Zahed Sheikh, the charity worker, and they began to build a case against him. He was still a mystery, but investigators caught occasional glimpses of his trail. He continued to live in Doha, the capital of

Qatar, a tiny nub of a country that juts into the Persian Gulf due east of Riyadh. Qatar was generally unknown in the West, but—after a coup in which a western-educated son supplanted his father as emir, it became one of the more liberal of the Gulf sheikdoms, and for that reason became an irritant to neighbors who had no interest in liberalization. At about the time Khalid Sheikh moved there, the new emir was preparing to launch Al Jazeera, the cable television channel that would pioneer a more open, Western style of journalism in the region. Doha was an odd place, full of new commercial office buildings and not much else. There was an old ramshackle downtown filled with Japanese and Chinese electronic goods. As elsewhere in the Gulf, a great many of the shopkeepers were Pakistanis, surrounding Khalid Sheikh with compatriots. He also had the advantage of a ready-made network of friends of his late brother, Abed Sheikh Mohammed, who had been killed in Afghanistan. Abed had attended university in Doha in the 1980s and became a much respected teacher in the city. He reformed a network of social clubs that had pre- viously been disreputable and made them a key feature in Doha's social and religious life.[16] Many people there had fond memories of Abed, who was further revered as a martyr of the Afghan war. He was still recalled fondly when Khalid Sheikh Mohammed arrived in 1993. Khalid was apparently unencumbered by his day job at the Ministry of Electricity and Water. Among his many travels during his Doha residency, he spent considerable time moving around the Persian Gulf, building and main- taining a fund-raising network.

"Throughout the region, there was this classic sort of money collec- tor—the guy who was hanging out at the mosque, checking out the scene, basically casing the mark, who would invariably be some old guy with lots of money. A religious guy, probably. The collector would come up along- side him, make his pitch very persistently, and the mark would write him a check,"[17] said one American official, who worked in the Gulf through- out the 1990s. "Khalid Sheikh Mohammed was a collector, a guy who would collect the money from the street collectors. He was in and out, operating in Qatar, UAE, Bahrain. The money network was very infor- mal. A guy in the Philippines would call a guy in Dubai who would call Khalid Sheikh Mohammed. It would be a chain of telephone calls, and

Khalid would send the money. . . . There were a very large volume of very small transactions involving several hundred people and he was in the middle of it. He's a mover of money. It was never clear if he was connected to anyone above, if there was a line going up. . . . Khalid Sheikh Mohammed would pop up in Dubai, Pakistan, occasionally Kuwait. He became an issue for me only because he was on my turf. So far as we could tell this was a bunch of individual actors."

Khaled Mahmoud, a Doha resident, recalled once running into Mohammed at a mosque in the city. They chatted for perhaps thirty minutes, during which they were repeatedly interrupted by people coming up to say hello to the short, plump, slightly balding young Mohammed, who was what American social-advice handbooks of the era would have called a very efficient and devoted networker. He was said to have a sharp sense of humor and an excellent memory. "He knew your name the second time you met him and remembered things about you from previous conversations," Mahmoud said.[18]

American understanding of new Islamist terrorism then remained inchoate. Attention and resources continued to flow toward actions aimed at potential state sponsors of terrorism, in particular to Iran, which had become the chief state proponent of fundamentalist Islam. American agents caught traces of Khalid Sheikh Mohammed all around the region and, indeed, the globe: Italy, Egypt, Singapore, Brazil, Jordan, Thailand, and back to the Philippines. In fact, he lived openly in the Gulf. "He wasn't even using an alias," said one American official who worked in the region then.[19] He often stayed as a guest at the estate of Abdullah ibn Khalid al-Thani, who was then the country's minister of religious affairs and who had originally urged Khalid Sheikh to move to Doha.

"Abdullah ibn Khalid had a farm outside Doha," said the American official.[20] "A lot of these guys had what were basically gentlemen's truck farms. It was a hobby. Grow cabbages, raise ducks. So he has this farm and he always had a lot of people around, the house was always overstaffed, a lot of unemployed Afghan Arabs. Khalid was always good for a handout. These guys would stay for while, quite a while, some of them; they came in looking pretty ragged and tended to get a little plumper after they'd been there for a while. You'd go to his house, there were always these guys hanging around and maybe a couple of Kalashnikovs in the corner, which didn't seem all that unusual. I'd been posted in Beirut, so having guns in the living room didn't seem to be that odd. A lot of these guys were probably honest mujahideen recuperating. It was not a bad gig."

American intelligence eventually figured out that one of the guys on the farm was the same Khalid Sheikh they had been chasing since Manila. They didn't yet know how significant his role had been. They thought Yousef had been the mastermind of the Manila plot and that his older uncle was little more than "a bad influence."[21] Nonetheless, a grand jury in New York issued a secret indictment of Khalid Sheikh Mohammed just after New Year's Day, 1996, and a debate ensued on what to do about it. FBI director Louis J. Freeh talked to Qatari officials, seeking assistance in arresting him for what was known as a "rendition" to the United States. Renditions were a relatively new tool then. They were effectively extra-legal extraditions, having the same effect—rendering an individual to a country that sought him—without having to go through host-country court proceedings.

One FBI official said months passed without the Qataris agreeing to arrest—or allow the United States to come and arrest—Khalid Sheikh, even though they acknowledged he was there. At one point, Qatar told the United States there was evidence that Mohammed was constructing a bomb.[22] They also said he possessed more than twenty different passports; still, they delayed approving the rendition. A meeting was called in Washington in early 1996. The meeting was chaired by Deputy National Security Adviser Sandy Berger, representing the White House, which was trying to build a consensus for going after Mohammed.[23]

Those who favored the rendition had imagined going in covertly with a team of perhaps twenty-five people. That was nowhere near the Pentagon's estimation of its needs. "We were off by orders of magnitude," said Jamie Gorelick, the Department of Justice's representative at the meeting.[24] Fearing another "Black Hawk Down" debacle, like the one in Somalia in 1993, the Pentagon insisted that a raid would require hundreds, if not thousands, of troops. This fear drove the Pentagon toward reluctance in many similar situations. The generals would typically say that if they were to attempt such operations they preferred to do so with overwhelming force. In effect, they proposed small-scale invasions, which they knew had virtually no chance of being approved. It was a way of saying no without having to say it. The CIA was similarly reluctant, if for other, less apparent reasons. Basically, in the mid-1990s they lacked both the capacity and the will to do that sort of thing without the military.

"Everyone around the table, for their own reasons, refused to go after someone who fundamentally threatened American interests," said another person who attended.[25] "The FBI can't go anywhere overseas without the CIA providing the intel, the DOD providing the logistics and military muscle in the event we have to shoot our way in. And none of that happened."

Absent both the backing of the Pentagon and the cooperation of Qatar, there was little that could persuade Berger to go in after Khalid Sheikh Mohammmed. It wasn't a question of desire. It was the purely pragmatic question of what could be done. Without military support, the answer was nothing.

In the end, rather than sending a kidnapping squad, FBI director Freeh sent a letter to Qatar's foreign minister. It said in part: "We believe, Your Excellency, that time is of the essence. I have recently received disturbing information suggesting that Mohammed has again escaped surveillance of your Security Services and that he appears to be aware of FBI interest in him. This information, together with his possession of false passports, leads to concerns that he may have already fled, or he is in the process of fleeing, Qatar for a safe haven. A failure to apprehend Khalid Shaykh Mohammed would allow him and other associates to continue to conduct terrorist operations."[26]

The letter had no apparent effect. In the end, when the Americans decided to send a very small team of agents to attempt a more judicious arrest, it was too little, too late. Mohammed was gone. "There may have been some FBI nudnicks on the ground in Doha, running around with handcuffs and some laminated card with Miranda rights on it," said one person.[27]

Different stories have been told about who exactly helped Mohammed escape. Most credit al-Thani, his host. One version has him fleeing to Prague, in the Czech Republic, which at the time was a thriving trading center for arms and information from the former Soviet Union.[28] In any case, it is less important where he went immediately than where he wound up—Afghanistan. There he renewed acquaintances with another newly returned wanderer: Osama Bin Laden. Sometime that year, 1996, Khalid Sheikh told Bin Laden he had an idea. In that it involved big buildings and airplanes, it was an amalgam of several prior terror plots: Yousef's attempted bombing of the World Trade Center; the Manila plot to attack several commercial airliners simultaneously; and Murad's idea to hijack an airplane and use the plane itself as a weapon, dive-bombing it into the CIA headquarters in Virginia.

Murad was a simple man. The Philippine report on his hijacking proposal conveyed the simplicity of his idea: "What subject have in his mind is that he will board any commercial aircraft pretending to be an ordinary passenger. Then he will hijack said aircraft, control its cockpit and dive it into CIA Headquarters. There will be no bomb or any other explosive that he will use in its execution. It is, simply a suicidal mission that he is very much willing to execute. That all he need is to be able to board the aircraft with a pistol so that he could execute the hijacking."[29]

Khalid Sheikh, however, was more complex, or at least a man who enjoyed the idea of complexity. He first mentioned Murad's idea to Bin Laden. Bin Laden said it was too simple. "Why use an ax when you can use a bulldozer?" he asked. Khalid Sheikh then proposed an outlandish plan to hijack ten aircraft at once, five each on the East and West Coasts of the United States. He and Yousef had dreamed about something like that. The two of them paged through almanacs of American skyscrapers, looking for targets. They considered, among others, Sears Tower in Chicago and Library Tower in Los Angeles.[30] Aside from the obvious symbolic value of attacking big buildings as emblems of American economic might, the skyscrapers had at least one other simple but useful quality as targets—they were easy to find. They were made so as to force people to pay attention to them.

Bin Laden and his associates had worked out what they regarded as the religious logic of attacking Americans years before. Mamdouh Salim, an Iraqi who was one of the original members of Bin Laden's shura, or ruling council, served several functions in the organization before his arrest in 1998. He was a trusted business executive, responsible, for example, for Al Qaeda's efforts in the 1990s to purchase materials to make nuclear weapons. He also served as one of Bin Laden's religious advisers. Bin Laden early on relied on others to determine what was proper Islamic behavior. Salim routinely issued fatwas for the organization. One of them concerned the advisability of attacking civilian targets: "Salim reasoned that if an attack on an American building killed Americans, that was beneficial. If Muslims working with the Americans were killed, these Muslims were being punished for helping the American infidels. If innocent Muslim bystanders were killed, they would become martyrs and go to paradise and would thus be grateful to those who carried out the bombing."[31]

Many members of Al Qaeda cells that have been broken up around the world in the last few years had been involved in low-level criminal activity, which they used to finance their terror plots. Some of Khalid Sheikh's own plots had been strapped for cash. This operation would be different: It was on a grander scale, more ambitious and expensive, and more closely controlled. It demanded men who were different, too— more cosmopolitan and technically proficient. With Yousef in jail, Mohammed didn't have his bomb maker, but crashing planes into buildings eliminated the need for one: the airplanes would become the bombs. What he would need instead were pilots. He asked Bin Laden for two things—money and martyrs. Bin Laden already had the money. The Afghan training camps he controlled would be a very good place to look for the other.[32]

BOOK THREE

The Plot

CHAPTER 1

The New Recruits

AFGHANISTAN

MARWAN AL-SHEHHI LEFT HAMBURG FIRST. His departure came as something of a surprise to friends. It was November 1999, and Shehhi was in his first semester after finally being admitted to Technical University Hamburg-Harburg. He'd had a rocky passage through what was normally a straightforward year of preparatory study. He'd made the year into three and finally got through it with passing marks. "He had just bought new furniture, books, he wanted to study and had asked me to help him with the math," said Mounir el-Motassadeq.[1] "He said he would bring his wife from the Emirates. And then he left."

Days later, Ziad Jarrah followed, then Mohamed el-Amir, and finally Omar, Ramzi bin al-Shibh. Jarrah and Amir, separately, took Turkish Airways flights, paid for with Shehhi's army scholarship money, through Istanbul to Karachi. It isn't clear what routes Shehhi and Omar took. Once in Karachi, the four men likely took domestic flights or buses to Quetta, the dusty desert capital of Baluchistan.[2] Al Qaeda had begun avoiding the Peshawar route whenever possible; it had become too well watched.[3] By the late 1990s, there was an almost paranoic fear of spies infiltrating the camps. Al Qaeda leaders had begun wearing masks to cover their faces, even in the camps.[4] Osama Bin Laden had succeeded perhaps too well in drawing the world's attention. His celebrity proved a double-edged sword—attracting danger as well as recruits—and necessitated more careful checking of newcomers. Some men were forced to wait at guesthouses on the Pakistan border for weeks before being brought inside Afghanistan. The men from Hamburg did not have that problem. They had been instructed to take taxicabs to the local Taliban office after their arrival in Quetta. Every driver will know where it is, they were told. And they did. They were told to leave everything behind at the Taliban guesthouse—all their belongings, passports, currency, even their names. As was standard procedure, they selected noms de guerre: Amir was Abu Abdul-Rahman, Jarrah chose Abu Tareq, Shehhi became Abul Qaqaa, and Omar picked Obeida al-Emerati.[5] Recruits were typically given local clothing and taken to the border, which was a barrier in name only. Sometimes recruits were taken on motorcycles on routes that easily circumvented the guard posts. Other times, they simply drove through the checkpoints without having to show any identification whatsoever.

Despite all that had gone on around and about them, life in the camps had not changed much for a decade. It was dominated, like most military life, by the humdrum of routine. Up early, down early, and in between, exercise, drills, and lectures, exercise, drills, and lectures. If it weren't for the prayers and the Afghan robes everyone wore, it would have been hard to tell it apart from basic military training anywhere in the world. The camp system, which at its high point encompassed more than fifty locations, under the supervision of various national fighting groups, was hierarchical. The main camps were sorting and marshaling yards where dueling jihadist groups competed for recruits. Khalden, between Kandahar and the Pakistan border, was the main Al Qaeda entry point. Men graduated from there to other camps, often passing through two, three, or even four locations, mainly in the hills and mountain regions of the southern provinces. They tended to be very spare operations with three or four small buildings for instruction and staff quarters. Recruits generally lived in tents. The best of them were offered specialized training in subjects such as sabotage, urban warfare, mountain climbing, rocketry, building cells, and building bombs. Even at the elite camps, ammunition and ordnance were in perpetually short supply.

A typical day would start with predawn prayers, then two hours of physical training—running, weight lifting, rifle practice. After breakfast, there was often an hour-long classroom session, then a nap, noon prayer, religious instruction, chores, the afternoon prayer, more physical training, evening prayer, and dinner. Recruits were usually in bed by 8:30.[6]

Bin Laden made regular circuits of the camps, giving lectures and encouragements. Some trainees reported meeting him several times. At one such meeting, Bin Laden showed the students results of a poll that indicated he was more popular in the world than the United States, which gave him great pleasure.

Even before Omar arrived to join them, Shehhi, Amir, and Jarrah were taken to the House of Ghamdi, an Al Qaeda facility near Bin Laden's Kandahar home. It was as if they had been summoned.[7] Waiting for them was Bin Laden himself.[8] The transformation from Hamburg students to willing warriors was about to be complete.

Khalid Sheikh Mohammed's plan to attack the United States had gone through several iterations since he first presented it to Bin Laden in 1996. It had changed in many ways, just as Al Qaeda itself had.

Afghanistan, in the years after the Soviet withdrawal, had been a riot of confusion. "A lot of the Arab youth didn't know what they were doing. When they were fighting in the civil war, they were sometimes told that they were still fighting the Russians," said Burhanuddin Rabbani, a Tajik political leader.[9]

The Taliban and Osama Bin Laden offered ways out of the confusion. Mullah Omar gave Afghans a vision of a unified society. Whether or not it was a society anybody else wanted, it would at least be clear about what it was. Bin Laden brought the same sort of focus to political Islam. Every Muslim country, it sometimes seemed, had at least one and often competing Islamist oppositions, each with separate complaints and preferred methods of operation. These did not suddenly disappear, but over a period of slightly more than two years—from the time he returned to Jalalabad in the spring of 1996 to the truck bombings of two American embassies in East Africa in the summer of 1998—Bin Laden became the popular public face of the cause, and not just by happenstance. He actively promoted himself and his ideas. From the very beginning Al Qaeda's *shura* had included a press officer. Like the other members of the organization, the press man had taken a nom de guerre. In his case, it couldn't have been more fitting: Abu Reuter, after the international news agency.[10] Reuter was effective, too. Bin Laden did numerous high-profile interviews with Arab, European, and American journalists and television programs.[11] He spoke softly but promised to bring the fury of the almighty down on the Saudis, the Americans, and the Jews. His was not a complicated program, but it aimed high, and partly because of this ambition it overwhelmed the clutter of competing Islamist causes.

Al Qaeda itself was never the huge organization its opponents sometimes portrayed. Its core was at most a couple hundred men, but they sat at the heart of a sprawling web of other like-minded organizations spread across the globe. At its peak, there were operational offshoots in sixty countries,[12] a network of networks that was able to take advantage of traditional Muslim customs of hospitality and aid to coreligionists. Al Qaeda sat at the center of this web but was never in any sense in control of it. American analysts sometimes likened it to the Ford Foundation of terrorism,[13] suggesting it sat back and waited for plans to be proposed, then accepted or denied them. This did happen, especially in the early years when it collaborated with other organizations rather than running operations on its own, but after Bin Laden returned to Afghanistan, he and his top deputies—his number two, Zawahiri, and military chief, Muhammed Atef, both not coincidentally Egyptians—transformed Al Qaeda into a small operational organization as well, that worked in cooperation or sometimes competition with other terror groups. Its own small size, which put severe limits on the skills available within its ranks, virtually required it to reach beyond its members for specific needs.

One underappreciated aspect of Al Qaeda operations was how crude many of them were. Intelligence analysts sometimes cited the plans' complexity and sophistication, as if blowing up buildings or boats or vehicles was high-end science. In fact, many Al Qaeda plots have been marked by the haphazardness of their design and execution. Over the years, many of the plots seemed harebrained at worst, ill-conceived at best, pursued by ill-equipped and unprepared, inept men. Some were almost comical in their haplessness: boats sank, cars crashed, bombs blew up too soon. Some of the men virtually delivered themselves to police. The gross ineptitude of the execution often disguised the gravity of the intent, and hid, also, the steadfastness of the plotters. Whatever else they did, they did not go away.

Al Qaeda took over several of the Afghan training camps previously run by different mujahideen factions and supported others financially. The number fluctuated over time. There were periods when, because of financial and security pressures, the group ran just a single camp.[14] Bin Laden's personal fortune has been the subject of great attention throughout his public life and it doubtless enabled him to begin his role as an underwriter of terrorism, but its size has routinely been exaggerated.[15] The Saudi government estimated he received approximately $1 million per year from the family from 1970 to 1994, when the payments ended. As important as the amount of Bin Laden's net worth was the fact that it gave him an entrée to wealthier segments of Saudi society and made him a great fund-raiser. Al Qaeda's own financial base was broadened far beyond Bin Laden's personal resources, which varied according to international financial constraints placed on it.[16] American and Saudi government funds that had financed the war against the Soviets were replaced by private fund-raising networks that operated throughout the world, some of them government-sanctioned charities that continue to operate.[17]

EAST AFRICA

While Bin Laden and most of his top aides spent the bulk of their time in Afghanistan consolidating, building, and tending the camps and their relationship with the Taliban, Khalid Sheikh Mohammed traveled the globe, searching out allies and recruits and assembling what later seemed a ubiquitous global network. "He was building a terrorism business," said Neil Herman, an FBI man who pursued Mohammed for years.[18]

Mohammed developed direct personal relationships with several of the men who became Al Qaeda's top regional operatives. His trail wound through European capitals, Africa, the Gulf, Central and Southeast Asia, and even Brazil.[19] Operating out of Karachi, where he had numerous residences, he made repeated visits to Southeast Asia—notably, Malaysia and the Philippines. He was said to be in Rome meeting with Tunisians and in Japan buying earth-moving equipment. Once, Philippine intelligence officials for the first time since foiling the Bojinka plot received information that Mohammed was back in Manila. The tip was specific, including information about where he would be and who he would be with.[20] He vanished before agents arrived to arrest him. At times, Mohammed would travel to other countries to personally establish terrorist cells and provide them with plans, money, manpower, and logistical support. Other times, he would operate at a higher level, overseeing local commanders who led their own cells.

During this period, U.S. investigators had no hint of Mohammed's deepening involvement with Al Qaeda. He didn't formally join the organization, and for years resisted pledging the traditional oath of allegiance to Bin Laden, but he was given direct charge of Al Qaeda's most ambitious plan, the American airplane plot.[21] The Americans continued to pursue Mohammed for his role in the Manila plot; that was sufficient cause to put a $2 million reward on his head, land him on the FBI's Most Wanted lists, and circulate Wanted posters around the world. In retrospect, it astonished investigators that despite the attention he remained so active.

"We reached out to every one of our friends out there to try and get him," said one senior Justice Department official.[22] "But he just kind of slipped off the screen."

Bin Laden had not. He was increasingly the principal focus of antiterror analysts. By the time he left Sudan in 1996 and relocated to Afghanistan, the intelligence community was mobilizing to stop him. The CIA established a Bin Laden station, the first ever agency unit focused on an individual, an acknowledgment that he represented a target the likes of which they had never seen before.

"We monitored his whereabouts and increased our knowledge about him and his organization with information from our own assets and from many foreign intelligence services. We were working hard on an aggressive program to disrupt his finances, degrade his ability to engage in terrorism, and, ultimately, to bring him to justice," said George Tenet, director of the CIA.[23]

If Bin Laden knew about the Americans' intent, he gave no hint of it. In February 1998, from a camp at Zawar Kili, near Khost in southern Afghanistan, Bin Laden issued a second declaration of war against the U.S. and announced a merger of Al Qaeda with organizations from

Pakistan, Egypt, and from across Africa. The new organization, which was little more than the old organization with a new public relations strategy, declared: "To kill Americans and their allies, both civil and military, is an individual duty of every Muslim who is able, in any country where this is possible."[24]

It was a call for a new generation of jihadis. Bin Laden held a news conference at the same camp not long after, repeating the same message in case anyone had missed it the first time. Behind him as he posed for the cameras was hung a map of Africa, an advertisement, in effect, for what would come soon after.

On August 7, 1998, trucks driven by Al Qaeda operatives rumbled through the midmorning traffic and crashed past security barriers at American embassies in Nairobi, Kenya, and Dar es Salaam, Tanzania. The bomb-laden trucks exploded in the embassy compounds, killing 224 people, injuring thousands, most of them Africans. Hundreds more would have been killed and hurt but for the extraordinary luck of there having been a filled water truck parked at just that moment in front of the Dar es Salaam embassy. It absorbed most of the blast, sparing untold damage. Still, the attacks were horrific both in their actual effects and what they portended. The attacks had been in the planning stages for years. Some of the participants had been sent to Africa many years earlier and patiently waited for the orders that came that summer. The simultaneity of the attacks amplified their impact and suggested a degree of sophistication and daring no one previously had granted Al Qaeda.

CIA director Tenet responded by declaring war on Bin Laden, either ignoring or oblivious to the fact that the war had been on from the other side for years. Nor did he know what form battles yet to come would take.

The airplane plot had already been two years in the making and had gone through several iterations. In the original plan, Mohammed himself was to pilot one of ten aircraft; rather than crash-land it into a building, he proposed slaughtering all the men aboard the airplane, contacting the media while in the air, then landing it in the United States, and making "a speech denouncing U.S. policies in the Middle East before releasing all of the women and children passengers." Bin Laden reacted coolly to this grandiose element of the plan and did not formally endorse any action until 1999, when the idea was slimmed down from the original ten flights to four.[25]

Bin Laden nominated four men to lead the hijackings—two Saudis and two Yemenis. The Saudis were Khalid al-Mihdhar and Nawaf al-Hazmi. The Yemenis were Walid Mohammed bin Attash and Abu Bara al-Yemeni. All were tough, battle-hardened Al Qaeda veterans who had fought in Bosnia or Chechnya. Attash, also known as Khallad, was a fairly senior Al Qaeda operative who had joined the jihad when he was just 15 years old, losing a leg in battle. He had for a time after the war supervised Bin Laden's personal bodyguard. For the last three years, he had played an active role in Al Qaeda's offensive operations. The early versions of the plan apparently did not contemplate these men training to become pilots. There was consideration of using other means of seizing control of the aircraft, even bombing them in midair as Yousef and Mohammed originally envisioned; early training for the hijackers concentrated on hand-to-hand combat.[26]

The would-be Yemeni hijackers applied for American visas early in 1999, suggesting the plot, like several others, might have been aimed at the millennium. The Saudis already had U.S. visas when they were chosen by Bin Laden. They had intended to attack the United States on their own, although it was never clear how.[27] The Yemenis were denied visas. There was no history of Yemeni terror against the United States and no reason to suspect the two men. They were denied because they were viewed as potential economic migrants, by far the most common reason applicants were kept out of the United States. With the denials, a new version of the plan was put forward. It proposed two hijacking crews in the United States led by the Saudis and two in Asia led by the Yemenis.[28] At this point, Mohammed involved his old friend Hambali to support the Asian portion of the attacks. Later, that plan was scrapped too, and the leadership council decided to go back to the essential elements of Mohammed's original plan—to hijack the planes by force and install their own pilots to fly them. That required, of course, finding pilots, or men capable of learning the job.

The Hamburg men gathered at the House of Ghamdi in Kandahar, near the start of Ramadan in December 1999. Bin Laden kept a residence at the Kandahar camp. In the past, he had used it for news conferences and interviews at which he proclaimed his jihad against the United States. He used it now to stage the next battle in that war.

Bin Laden asked each of the Hamburg men if they would join ranks with him, pledge their loyalty, and—harder still—accept suicide assignments. Becoming martyrs in a holy war was something they had talked about for years. Some of them had long dreamed of the opportunity. They had debated the morality of martyrdom, had talked of how dying for Islam differentiated the death from suicide, which was expressly forbidden. Martyrs, they believed, would reach the highest level of paradise.[29] There had been periods back in Harburg when they talked about these issues every day for hours. What was their responsibility, not to Bin Laden or Al Qaeda, but to God?

Amir had said not long before that Bin Laden might or might not have the right answers, might or might not be pursuing the right course. He had just months before warned one of his study-group members to keep himself strong, to follow the path of righteousness and steer clear of radical groups.[30] Just before coming to Kandahar, he had asked his mother if he could move back to Egypt to care for her.[31] It was almost as if he were asking someone else to stop him from something he knew he couldn't stop himself from doing.

Ziad Jarrah would have to give up something the rest of them knew not the least bit about—Aysel and everything she represented. Not just a girlfriend or a wife but the very real knowledge that this was someone he could spend the rest of his life with, someone who transformed the idea of that mortal life into something beyond just himself.[32]

Omar and Marwan al-Shehhi seemed to harbor the fewest doubts. Omar had his dreamy acceptance of the word of Allah, his almost mystical Yemeni belief in the goodness of Islam. Shehhi had been eager from the moment he arrived in Harburg. He treated martyrdom not as a subject of morbid fate, but rather of joyful embrace. He celebrated its prospect, danced and sang songs about it. Maybe that happy countenance masked an iron resolve; maybe it was Shehhi's embrace of the prospect that had been missing, the thing that pushed the group to the answer its members gave in Kandahar, which answer, of course, was yes. They accepted Bin Laden's mission.[33]

Shehhi, who had fallen ill, left not long after and flew home to the Emirates. He immediately applied for a new passport and a U.S. visa. On

January 3, 2000, he ordered a pilot-training video online, the first concrete step toward flying in the entire plot.

Meanwhile, Bin Laden sent Amir, Omar, and Jarrah to his military commander, Mohammed Atef, who gave them the broad outline of the airplane operation. Amir met several more times with Bin Laden and was selected to lead the group. The three men then left for Karachi, where Khalid Sheikh Mohammed was waiting with more specifics of the assignment. Mohammed told the young men they should return to Germany and find training that would prepare them to fly commercial airliners.

They agreed and departed as they had come, one by one. Amir and Jarrah headed back to Germany. Omar took more time; he returned to Afghanistan and didn't resurface in Europe until March in the Netherlands, just days after he had recorded a martyrdom video in Kandahar.[34]

KUALA LUMPUR

There must have been a sense of frustration then within Al Qaeda. Weeks before, a plot to bomb several targets in the United States and elsewhere at the dawn of the millennium had been broken up. One man was captured as he entered the United States from Canada. Other portions of the operation were disrupted in Jordan, and an attempt to torpedo an American warship in Yemen foundered. The ship, USS *The Sullivans*, had been anchored in the harbor at Aden, on the southwestern coast. The would-be attackers had reconnoitered just up the rocky shoreline where they loaded a small boat with a huge amount of explosive, intending to ram into the side of *The Sullivans*, detonate the explosives, and, they hoped, sink the ship. Instead, their own overloaded boat was swamped and nearly sank just as they put it into the water. They were able to rescue the scow, but the mission was scrubbed.

After those failures, Bin Laden must have been reluctant to scrap any portion of what the leadership had come to call "the planes operation." For the time being, they proceeded with both the newly found German pilot candidates and the original four jihadis Bin Laden had chosen to lead the attacks—the Saudis, Mihdhar and Hazmi, and the Yemenis, Attash and

Bara. The four gathered for a meeting in Malaysia that Hambali, Khalid Sheik Mohammed's old comrade from the Afghan war, would host.

Hambali had eluded detection altogether when Khalid Sheikh's Manila air plot unraveled. No one even knew he was involved. Filipino investigators eventually found the link to the front company he had established in Kuala Lumpur, but no one associated Hambali with it. He had remained in his little hut along Manggis River Village Road and gone to work building a local, then a regional, terror network. It was an unlikely place from which to command anything, about as far off the beaten path of world affairs as one could get. Yet Hambali and two other exiles from Indonesia sat in their tiny Malaysian village and meticulously planned, then patiently built an extraordinarily disciplined network, Jemaah Islamiyah, that could serve as a model for anyone anywhere who wanted to do the same. It had more structure than anything Bin Laden ever attempted, with strict geographic sectors that covered all of Southeast Asia, a command table of organization in each of the sectors—which they called *mantiqis*—with clear lines of authority and responsibility up and down.[35] Unlike Arab terror organizations, which tended to be haphazard at best, more commonly ad hoc, Jemaah Islamiyah was strictly organized. Its regional commanders held regular meetings and assessed monthly taxes on members. They might have been squatting in the red clay mud in the middle of a disappearing rain forest, but Hambali and his two partners were far from lacking sophistication. Together, the three embarked on a long, patient recruiting process. The two older men preached frequently at local mosques while Hambali sat quietly in the back. He was careful. He spoke only to small groups and in private. One man said what was most impressive about Hambali was not his intellectual prowess, which was evident, but "his quiet and humble manners,"[36] not characteristics usually associated with terror masterminds. He was all the more dangerous because of it.

Hambali gathered money and men. He recruited a remarkably diverse group—petty traders, artisans, and factory workers, but also engineers, businessmen, middle managers, and university lecturers. They found Hambali to be a man of tremendous, if quiet charisma. "He was such an unassuming person," said Mohamad Sobri, a former soldier in the Malaysian army who became one of Hambali's followers. Sobri allowed his house to be used as a meeting place. As many as thirty men met every month to discuss Islam and its place in the world. Except for the geography, the talk was remarkably similar to that in Hamburg.

"We talked about Chechnya, Afghanistan, Bosnia, Palestine, and the sufferings of Muslims in [the Indonesian islands of] Ambon and Maluku," said Sobri.[37]

Hambali made a regular circuit of such small prayer groups, promoting a goal—to unite the Muslims of Southeast Asia into a single, powerful Islamic force. With nearly 300 million Muslims throughout the region, it would be by far the biggest Muslim nation on Earth. Hambali promoted this goal and a means to attain it—jihad. At the end of every session, followers passed a hat around the room—at their own, not Hambali's, suggestion—and whatever small amounts of money they could spare, they gave to him. He recruited men to go to the camps in Afghanistan, as he had done as a young man.[38] He enlisted others in small organizational tasks. It was at the Afghan camps that he had met Khalid Sheikh Mohammed. Both had trained with Abdur Rasul Sayyaf. Neither forgot the other. Mohammed brought Hambali into the Manila air plot to help launder money and provide logistical support. Hambali's network helped Ramzi Yousef, who was at the time among the most wanted men on Earth, enter the Philippines surreptitiously by the well-worn smuggling route through the southern islands. The back door, they called it.

Few intelligence agencies were paying much attention to JI. Mohammed was watching, however, and asked Hambali to bring his fledgling group in with Al Qaeda. Mohammed, accustomed by now to the largess of Gulf oil sheikhs, was impressed with how much Hambali had done with relatively meager sums of money.[39] He brought him to Afghanistan to meet Bin Laden, who was similarly impressed. They agreed to join forces on targets of mutual interest, and Bin Laden began providing funding.[40] When Bin Laden's Yemeni hijackers were refused visas to the United States and the plot was reshaped to include Asia, Mohammed sought Hambali's assistance once again, and he agreed to provide what assistance he could.

Kuala Lumpur was chosen for the meeting largely as a matter of convenience: The Yemenis intended to study airline security operations in

Southeast Asia, for their portion of the operation, anyway. Attash, who had lost a leg in the Afghan war, also wanted to visit a clinic that had provided numerous jihadis with prosthetic limbs. The two Saudis were en route to the United States to become pilots and Malaysia was a suitable way station. Finally, Muslims could enter the country without visas. Since Mihdhar had missed the training sessions in Afghanistan, this would be the only time the four key field operatives would be together.[41]

Unbeknown to Al Qaeda, both Mihdhar and Nawaf al-Hazmi were being tracked by Western intelligence services as they traveled for the meeting. The month before, Mihdhar had been in Sana'a, the capital of Yemen. Mihdhar, a Saudi national, was related by birth and marriage to two Yemeni clans renowned for their bellicose nature. One of them, the Hada clan to which his wife belonged, was an especially fierce group, who, in addition to other local activities, had long been involved with Bin Laden agents. For several years Bin Laden had used a telephone at a Hada home in Sana'a as a sort of switchboard, a place to call for messages to be transferred in and out of Afghanistan. The number had been revealed in previous investigations, and all traffic to and from it was intercepted. That autumn, calls to the number indicated Khalid, Nawaf, and Salem were intending to meet with other Al Qaeda agents in Kuala Lumpur.

Intelligence agents could establish only tentative identifications of the men named in the calls, but Mihdhar was tracked as he traveled to Malaysia. A photocopy of his passport was obtained as he transited through Dubai in the United Arab Emirates; the copy was provided to the CIA before Mihdhar ever landed at Kuala Lumpur International. The Americans were also able to positively identify Nawaf al-Hazmi before he arrived. At the CIA's request, Malaysian domestic security picked up the trails of the two men, tracking them to a local hotel and subsequently to the Bandar Sungai Long condominium complex outside the city.[42]

One of the units in the complex belonged to a Hambali deputy, Yazid Sufaat, a former Malaysian army captain and Cal State-Sacramento biochemistry graduate. Sufaat had been sent on a government scholarship to the United States for education, as a great many talented East Asians were, as an investment in Malaysia's future. Sufaat and his wife, also a Cal State alum, had prospered after their return to Kuala Lumpur. She owned a computer services firm that mainly did back-office support work; he owned a company that did personnel drug testing for the government, a lucrative business in a rigid, authoritarian society determined to deny Western vice a foothold. They and their young children lived in a modest two-story row house, with an older Mercedes sedan and a Proton compact in the drive. It was not lavish; the house had the corroded look of many things in the tropics, where time, heat, and humidity conquer all; any exposed metal was rusted, the paint peeled. But the couple were able to buy a first-floor condo as a weekend getaway at Bandar Sungai Long, a new development in the hills out of town. The complex advertised "city living, country style." You could buy units "*siap* and *sedia*," prefurnished top to bottom with Russian pine furniture from Ikea. With its Jack Nicklaus–designed golf course, sports clubs, foot reflexology, and post-partum slimming classes, the development could have been in any southern California suburb. One notable difference was that Sufaat sometimes lent the condo to Afghan war veterans who came to town to be treated at a prosthetics clinic. It probably didn't seem all that odd then when, just after New Year's, 2000, a small group of Arabs, one missing a leg, showed up at the condominium.[43]

The men used the condo—coming and going—for three days. They left for shopping excursions in Kuala Lumpur. Several were photographed having poolside phone conversations at the community recreation area outside. It was not until ten months later that CIA analysts figured out who most of the men in the photos were, not until Attash was identified after the fact as one of the operational commanders of the attack on the USS *Cole* in Aden.

On January 8, the men left Kuala Lumpur. Attash, Mihdhar, and Hazmi departed Malaysia through Bangkok, although the security services lost track of them and did not determine they had gone to Thailand until later. Attash went to Bangkok to meet two other operatives. Mihdhar and Hazmi went because they thought it a less suspicious point of departure for the United States.[44] On January 15, the two Saudis, traveling under their own names, undetected, flew out of Bangkok to Los Angeles.[45]

CHAPTER 2

Preparations

BOCHUM, GERMANY

AFTER ZIAD JARRAH LEFT HAMBURG in late November 1999, a friend called Aysel Sengün at her flat in Bochum, where she was continuing her studies in dentistry. The friend warned that he had heard rumors about Jarrah, that he was up to something, that he might have gone to Afghanistan. The friend said Jarrah's family in Beirut had called and was frantic with worry. They wanted to know where he was. Of course, Aysel didn't know either.[1] Jarrah had told her he was going to Beirut, which, even as he said it she hadn't really believed, but to have the lie confirmed so starkly took her aback. What had he told his family, she wondered? Nothing, as it happened. He hadn't said a word. She began calling around Hamburg to find out if anyone knew where he was. Just as she had before when she couldn't get hold of him, she called every number she could think of and then some. The flurry of phone calls tripped a wire somewhere in the Hamburg network. Somebody got worried and passed the worries up the line. Within a few days, Omar, who had left Hamburg for Afghanistan but was making intermediate stops in the Netherlands and Yemen, heard about Aysel's questions. Jarrah's closest friends in Hamburg had been among the Moroccans, so Omar called Mounir el-Motassadeq and asked him to call Aysel and calm her down.[2] Motassadeq did as he was asked. He phoned Aysel and tried to reassure her. Jarrah was fine, he said. No, he didn't know where he was, but he was sure he'd be home soon. Don't worry.

The call did little to assuage Aysel. She didn't know any more than she had known before, only that people she didn't know were calling her and telling her not to worry, which in itself was troubling. More important, she still didn't know where Jarrah was, who he was with, what he was doing. Then, a couple of weeks later, a letter arrived in Aysel's mail. It bore a Yemeni postmark and handwriting she didn't recognize on the envelope. Inside, however, was a handwritten letter from Jarrah. He was well, he said. Don't worry, please. I will be back soon. Jarrah told her that when he came back he wanted them to have a child together. "The special thing," Aysel said, "about the word 'child' was that he wrote it in several different languages. He also wrote that he missed me."

She was, of course, ecstatic.

"I was incredibly happy with that letter because now I knew he was alive," she said. "I told his parents at once I had received a sign of life."

A week later, Jarrah called.

"He told me he would be home soon. But I can't recall the exact content of the conversation because I was so excited," she said.

Jarrah's trip back to Bochum was not without incident. After being told he was to go to the United States for flight training, he left Khalid Sheikh Mohammed in Karachi and flew to Dubai to change planes for Germany. Immigration officials noticed he had a photocopied page of the Qur'an attached to his passport and they stopped him for questioning. They found his luggage full of religious materials—cassette tapes and pamphlets. His itinerary, too, aroused interest. Any young Muslim men who had been to Pakistan for vacation or schooling, the typical claims jihadis made, were suspect, especially if they had remained in Pakistan, according to entry and exit visas, for an extended period, as Jarrah had done. They pulled Jarrah aside.[3]

It had been just hours since Jarrah left Mohammed. He must have still been in a daze as the import of what he had agreed to do sunk in. He talked freely with the Emirati security officers. He told them he had been to Afghanistan and was going to go to the United States, to preach Islam and learn to fly airplanes. The officers interrogated Jarrah for several hours; meanwhile, they contacted American officials and asked what they would like done with this very suspicious character.

"What happened was we called the Americans," said a UAE official. "We said, 'We have this guy. What should we do with him?' . . . He's there, we called the Americans, their answer was, 'Let him go, we'll track him.' We were going to make him stay. They told us to let him go. We weren't feeling very happy in letting him go."

Jarrah was released. He went shopping in the mall incorporated into the Dubai airport; he bought gifts for Aysel, then boarded his flight to Germany. He spent a couple of days in Hamburg. While there, he reported his passport lost. This was standard practice for jihadis who were planning to go to the United States, a way of erasing previous travel. Then, suddenly, without warning, on a day in early February, there he was, standing at Aysel's door: Clean shaven, neatly dressed—the Ziad she knew from their first weeks in Greifswald, the Ziad she had fallen in love with. He gave her the gifts—jewelry, honey, shoes, a skirt.

"And of course I asked the question, 'Where have you been?' And I did not ask it once. I asked it a lot of times. The only answer I got was, 'Don't ask me.' Later he would say, 'Don't ask me, it's better for you.' That sort of irritated me, so I asked, 'Why was it better for me?' I would not receive an answer.

"At some point, I just told myself, 'It's OK,' and I was content with the situation. Basically I was happy that he was here and that his *sturm und drang*—that's how I interpreted this time—was over. . . . He came back without a beard and I hoped that he had decided for me."

That night, as Jarrah lay asleep in her bed, Aysel lifted the blankets and carefully examined his body, top to bottom, every inch of it, looking for bruises, scars, anything that might tell her where he had been and what he had done. There were none. He looked fit, athletic. He's OK, Aysel thought. She said later she made a choice then to accept his lack of explanations. He was back and that was all that mattered. Whatever had happened was in the past. Everything would be like it was in the beginning. For a while, it was. Jarrah seemed more relaxed about his religion, more moderate. He told her a little about Pakistan, but mainly about the landscape and how differently and simply people lived. He never said a word about Afghanistan or why he'd gone. He told Aysel that while he was away he had come to a decision. He knew now what he would do with his life: he wanted to become a pilot. It made sense to Aysel. He had told her years before how much he had loved flying as a child; he constantly drew pictures of airplanes.

Immediately, Aysel began making plans, imagining their life together in the kind of loving detail that made her eager to get on with it: another year in Germany for Ziad's training, then children, maybe a year or two in Turkey with her family. If Ziad wanted to leave Germany, he could work for the Turkish airlines and she could work as a dentist. Then they could move on to Beirut if he wanted. Everything seemed possible. The two of them together plotted and talked, dreaming just as they had in Greifswald. Only now, instead of plotting to escape the gray dreariness of eastern Germany, instead of searching for new universities, they looked for flight classes. They contacted local flight schools, but there weren't many near Bochum. Ziad took off on his own some days. He visited a cousin he hadn't seen in years who worked as an engineer at a nuclear power plant. He went to visit friends in Hamburg, Berlin.

One day, when Ziad was out, Aysel came home from school to find a message for Ziad on the answering machine. It was from Munich, from the German representatives of a flight school in Florida. The message indicated they were responding to Ziad's query. Aysel was furious. This was more like the old Ziad than she had bargained for—lying, hiding information. She confronted him when he came home. As usual, Ziad had an explanation: It's the best training, he said, and, more to the point, the fastest. He could earn his license in the United States faster than anywhere else and it was the best license to have. You could transfer it anywhere. Besides, he said, I have to get away from my old friends. This is the only way to do it. Aysel was mollified, at least as much as she ever was.

"Before he left for Pakistan, he made it clear he didn't want to live in Germany anymore, but when he came back and didn't want to talk about the time of his absence, I thought he did something wrong, which was probably not legal because even at that time the Chechnya war was going on. And I thought because of his religious beliefs he sort of did his share in this thing. That's what I meant by *sturm und drang*. . . . When he came back without a beard to Germany, I thought his religious intensity has gone softer. And is more compatible with my Western lifestyle."[4]

She put her doubts aside and acquiesced. Not long after, she e-mailed a friend: "I know he did some bullshit. . . . I can't quite believe it, but from now on he can't say anything about what I do or how I behave. I know more than he thinks I know."[5]

In March, Jarrah went down to Munich, signed a contract and paid a deposit for commercial training from Florida Flight Training Center in Venice, Florida. He would start in the summer. What Aysel didn't know was that some of Jarrah's mysterious friends—the men she never met because, as he told her, women couldn't go where he went—would be in the United States with him. Aysel also didn't know that Jarrah had earlier told a cousin he thought it would be great to be a Muslim martyr and that a plot had been set in motion to achieve exactly that end.

SAN DIEGO

By the time Western intelligence agents determined they had been in Bangkok, Khalid al-Mihdhar and Nawaf al-Hazmi were gone, on board a United Airlines flight to Los Angeles. They arrived January 15, 2000, and spent the next couple weeks in the city.[6]

Early in February, while they were eating lunch at a restaurant in Culver City, outside Los Angeles, a man named Omar al-Bayoumi happened to walk by their table. Bayoumi dropped a newspaper on the floor, bent to retrieve it, then told the two men at the table he couldn't help overhearing them speaking Arabic.[7] He struck up a conversation with them and soon was inviting them down to San Diego, where he told them he could help them get established. This was the sort of thing Bayoumi did often. He acted as a sort of self-appointed Arab welcome wagon for the greater San Diego metropolitan area. He seemed to know every new Muslim who came to town as soon as he arrived. Or, as in the case of Mihdhar and Hazmi, before they arrived.

It was never clear to the people who met Omar al-Bayoumi in San Diego who he was, whom he worked for, why he came to California, or why he left. Bayoumi arrived in San Diego in 1995. He lived with his wife and four children at a suburban apartment complex. He told people he was a student of international business, but it seemed unlikely because he was already 40 years old and, more to the point, he never went to school. He didn't work, either. At least not at a job. He explained that to those who asked by telling them he received a monthly stipend from his former employer, Dallah Avco, an aviation company in his native Saudi Arabia. He told others he was on a Saudi government scholarship.[8] Bayoumi carried a video camera with him almost everywhere he went; he taped everything from local soccer matches to casual dinner gatherings to sermons at the Islamic Center of San Diego, which was the hub of the city's multiethnic, 100,000-strong Muslim population. He paid close attention to newcomers and could be counted on to help them find housing and, in general, get settled into American life. He had good connections at the Saudi consulate in Los Angeles, particularly with the religious affairs officer there.[9] The Saudi government, like many Arab governments, keeps track of its citizens when they're abroad. They pay particular attention to young men, trying to ensure they do nothing to embarrass the kingdom. Many people assumed Bayoumi was some sort of minder paid by the government to do exactly that. It was seen as a largely benign activity.

In February, just after Ramadan, Bayoumi brought Mihdhar and Hazmi to town and asked people to help them settle in. They hardly spoke English and would need help getting Social Security cards, driver's licenses, and bank accounts. Bayoumi put them up at his place for a few days then brought the two men to the two-story Parkwood Apartments in Clairemont, a pleasant, middle-class suburban neighborhood north of the central city.[10] Bayoumi arranged an apartment and even loaned them money for their security deposit and first month's rent. He threw them a welcome party. Bayoumi told people they were in San Diego to learn English, although no one, including Bayoumi, could remember either of them ever going to a single class. Hazmi did spend a lot of time, a friend said, at San Diego State University. He used the college library computers for hours at a stretch, surfing the Web and talking in online Arab chat rooms. Hazmi's English was spotty, but he was an easygoing, sweet-natured man who got on well with almost everybody. In the mornings, he'd stop by the apartment rental office and say hello to the managers, staying for coffee and cookies. He was cleanshaven, small—some guessed his height as less than 5-foot-6—and thin, a homebody who stuck close to his room when he wasn't at the university.

Mihdhar was tougher to read, in large part because he spoke no English whatsoever. The two men had known one another since childhood in Mecca, Saudi Arabia. They talked often about the war in Chechnya, the prior war in Afghanistan. Hazmi told one man that to fight for Islam would be "a big honor," but never mentioned that the pair of them had fought in Bosnia just a few years before. Hazmi also expressed admiration for Osama Bin Laden, who he said was acting on behalf of all Muslims.[11]

Hazmi signed a six-month lease on the ground-floor Parkwood unit, but he and Mihdhar complained that they couldn't afford the place. After a couple of months, they sublet to another man—also steered their way by Bayoumi—and moved out, taking a room in the house of a retired literature professor they had met at the Islamic Center, Abdussattar Shaikh. The home was in a rural residential part of east San Diego County, perched on a bluff overlooking Spring Valley. The large two-story house in the sleepy Lemon Grove neighborhood has been a gathering spot for Middle Eastern men for two decades. The meetings aroused more curiosity than suspicion. Shaikh was prominent in local civic and Islamic affairs. He often rented rooms to young Arab men and frequently hosted Qur'an classes. The classes and the meetings served a double purpose. Shaikh helped the young men, and he profited from what he learned from them. He was an informer for the FBI.[12]

In the spring, Hazmi and Mihdhar told friends they were interested in taking flying lessons. They asked around and when they discovered how much it would cost, Hazmi told a friend he could have money for the lessons wired from Saudi Arabia, but he didn't have a bank account. He wondered if the friend would allow $5,000 to be wired to his account.[13] The friend agreed, but when the money arrived in mid-April it was from the United Arab Emirates, not Saudi Arabia, and the sender was identified only as Ali, which spooked the friend a bit.[14] He told Hazmi he would not help him again.

As it turned out, there was no need. Soon after, another friend took Hazmi and Mihdhar to nearby Montgomery Field, a small civil aviation airfield nearby. They signed up for a series of lessons.

"The first day they came in here, they said they want to fly Boeings," said Fereidoun "Fred" Sorbi, the instructor. "We said you have to start slower. You can't just jump right into Boeings."

Sorbi gave them introductory lessons in one of his small Piper Cherokees, each taking a turn at the controls for about an hour. "We took them up to show them how the airplane flies," said Sorbi. On approach to the runway at the end of the flight, Sorbi said, one of the men—he couldn't remember which—grew extremely frightened and started praying aloud, calling out to Allah as the other man piloted the small plane toward the landing strip. Sorbi advised them to delay further lessons: "We told them to go to college and learn to speak English if they wanted to become pilots. They said they were."

That was the end of the flight lessons. Not long after, Mihdhar left town for good. He told acquaintances he had family problems at home in Yemen.[15] On June 15, he flew Lufthansa to Frankfurt, continuing to Oman.

Hazmi told Shaikh he still wanted to become a pilot. He also told him he wanted to take a Mexican bride. "So I taught him a few Spanish phrases, like *que pasa*," Shaikh said.[16] Hazmi even posted a message on a lonely-hearts website: "Saudi businessman looking for a bride who would like to live in this country and Saudi Arabia." Shaikh said there were only two responses, both from Egyptian women.

"While he lived with me, I never saw him use a telephone. I wondered if he had any family at all," Shaikh said. "He told me once that his father had tried to kill him when he was a child. He never told me why, but he had a long knife scar on his forearm. . . . He said he came here to learn English, but I didn't see him going to school very often. He told me he was taking English classes at a downtown language school."[17]

At times, Hazmi seemed almost to be on a sort of permanent vacation. He bought season passes to Sea World and the San Diego Zoo. He also bought a blue Toyota sedan and frequently made the run up through the Mojave to Vegas. In town, he hung out at Cheetah's, a nude bar near the Islamic Center. The center itself was hardly a haven for radical Islam. It was multiethnic and promoted assimilation. All the signs in the building were in English. When, later in the year, a group of men showed up and passed out leaflets praising Bin Laden, center officials confiscated the literature and told the men to leave and not come back.

Hazmi took a job for a short while, washing cars at a Texaco station. The station was owned by two Palestinians and served as a hangout for Arab men, who sat outside at a picnic table, talking and sipping coffee. Hazmi came by the station and hung out at the picnic table even when he wasn't working. Other than the few weeks he worked washing cars, for the eleven months he lived in San Diego he had no evident source of income.[18]

FLORIDA

Mohamed el-Amir did not return to Hamburg from Afghanistan until the end of February 2000.[19] He didn't bother to rent a new apartment; his friends arranged temporary rooms at student dormitories or loaned him their apartments. There was no need to make more permanent arrangements. Amir immediately began contacting flight schools in the United States. In March, he e-mailed thirty-one different schools. He wrote: "We are a small group of young man (2–3 persons) from several different Arab countries. We would like to start a course for professional airplane-pilots."[20] Amir also e-mailed a friend from his university days in Cairo who was living in Florida, asking if he needed to apply for a student visa before coming to the United States.[21] "It is a quite difficult period, but also quite interesting," he wrote.

The entire group that had gone to Afghanistan applied for new passports, shedding the record of travel to Pakistan. Shehhi obtained his new papers in the Emirates before he ever returned to Germany. Omar had gotten a new passport in Yemen even earlier and Amir, like Jarrah, received one when he returned to Germany. All said they had lost the old ones. On January 18, Shehhi, using his brand-new passport, became the first of the men to receive a U.S. visa. Amir and Omar went to Berlin and applied for theirs. Amir's request was approved without incident. Omar's was not. It was the same problem experienced by the original Yemenis Bin Laden chose for the hijacking plot—consular officials were predisposed to deny Yemeni visa requests. Omar was interviewed and judged to have insufficient ties to Germany and was seen, therefore, as a likely economic migrant and denied.[22] Jarrah received his visa, like Amir, without incident, in May.

Two other members of the Hamburg group, both Moroccans, followed the others to Afghanistan: Zakariya Essabar and Mounir el-Motassadeq. Essabar left quickly with little warning in January. This was not out of character. He was the man who had become, two years earlier, the instant radical, the one who within a few months went from being a haphazard college student to a rabid fundamentalist; he was capable of cutting off friendships in the space of weeks, to become a religious enforcer. Motassadeq was more hesitant. He had been a fervent fundamentalist for years, so much so that German authorities had noticed and monitored his activities; they put his name on border watch lists, marking him as someone whose travel ought to be observed.[23] But he had never shown any intention of acting on his religious impulses. After Amir returned from the camps, Motassadeq quizzed him about the experience. He saw Shehhi and Omar, too, and had similar conversations.[24] Unlike many of the other men, Motassadeq had kept in regular contact with his family back home and involved them in his decisions. He was a solid student, not gifted but steadfast. He married a Russian émigrée and sought the blessing of his family for it. Before he left for Afghanistan, he and his new wife flew home to Morocco. He didn't mention Afghanistan, but told his father he wanted to go to Pakistan. Not only did his father approve, he gave him $2,500 to finance it. Motassadeq returned to Hamburg, then took off for Istanbul on May 22.[25]

A week later, on May 29, Shehhi flew from Brussels, Belgium, into Newark International Airport in New Jersey. He checked into a New York City Courtyard Marriott the next day, then moved to a Best Western while he waited for Amir, who followed five days later, flying out of Prague to Newark. Their detours through third countries were intended to make it harder to trace where they had gone.[26] Both Amir and Shehhi were admitted for six-month visits. Amir's visa was issued using the first and last names in his passport. As is often the case with Arabs, neither Amir nor anyone in his family had ever used that last name. In effect, then, when he came to the United States he came as a brand-new man—Mohamed Atta.[27]

Atta and Shehhi spent their first weeks in the United States in New York. If they went anywhere, Shehhi at least was back in the city toward the end of the month when he received a wire transfer of $2,000 from Omar in Germany. Atta purchased a cell phone and a service contract, giving a Norman, Oklahoma, address for billing, and in early July, that's where the two of them headed.[28] Although Atta had contacted dozens of flight schools from Germany, he had not made arrangements to attend any of them. He seemed to want to inspect them personally. He and Shehhi spent three days in Norman, visiting the Airman Flight School there, but something evidently was not to their liking.[29] Within the week they were on the Florida gulf coast, setting up housekeeping and bank accounts and enrolling at Huffman Aviation for commercial aviation pilot training courses. By the time they arrived, Ziad Jarrah was already taking classes a couple hundred feet away at the Florida Flight Training Center. Both schools were located at the Venice Airport. Jarrah took a room by himself at a small apartment building less than a block from the airport. It would have been hard for the three Arabs not to run into one another at the little airport in the little town, but there is no record of anyone seeing them together.

Atta and Shehhi stayed for a week in a rented room at the home of their school's bookkeeper, but it quickly became apparent the arrangement wasn't going to work. The boarders were simply too messy. As their training got underway, they rented, for $550 a month, a small pink stucco house in Nokomis, the next town north, where they stayed for several months.[30] They brought the rental agent cookies when they moved in. The agent was impressed with their frugality. In the suffocating heat and humidity of Florida in summer, they seldom used the air conditioning.[31]

The two locations were separated by a continent, but the central Florida coast where the men from Germany lodged had something in common with San Diego County where their Saudi counterparts landed. California and Florida, each blessed with abundant sunshine, were the top two states in the nation in numbers of pilot training institutions. Flight schools were so numerous in Florida that the state sometimes called itself the "aviation state." Both places have fast-changing and booming populations. San Diego, in particular, had a vibrant Muslim community that welcomed Mihdhar and Hazmi. The tiny towns in west Florida where Atta, Shehhi, and Jarrah studied and lived had virtually no Muslims and fewer

Arabs. The resident populations of the towns were overwhelmingly white and elderly, and the two men, by design, made many fewer local connections than Hazmi had in California.[32] Where Hazmi and Mihdhar gave up on flight training after a couple ground schools and introductory flights, Atta, Shehhi, and Jarrah were solid students, logging flight hours as quickly as they could. They even went to flight schools and simulators in neighboring communities for extra training. In the case of Atta and Shehhi, the resolve had to come from Atta. Shehhi had never been much of a student in Germany, while Atta ground away at whatever he did. Diligence was his hallmark.

Atta and Shehhi's Florida training had few distractions. Once they were settled in, they began receiving a steady stream of wire transfers into their local bank account. The transfers, which in four months would total more than $120,000, included one single transfer for $70,000, from the United Arab Emirates. The usual sender there was another of Khalid Sheikh Mohammed's nephews, Ali Abdul Aziz Ali.[33] Ordinary wire transfers, completely legal and typically unexamined—the same mechanism, for example, that mortgage lenders often use—were the principal financial mechanism supporting the pilots in the United States. In addition to the UAE money, they also received lesser amounts from Omar in Germany. That money was drawn from Shehhi's German bank, where his UAE Army stipend was deposited. The German transfers occurred at times when money was slow to arrive from Ali.

Atta and Shehhi took their classes, did their flying and, judging from bank records, shopped for most of their food and—with a couple of exceptions at a Sarasota mall—everything else at local Wal-Marts.[34] They were well enough liked by classmates, who invariably described them in much the same way casual acquaintances described them in Germany: Atta was brusque, formal; Shehhi was smiling and amiable. They were seldom seen apart and were so close many people assumed they were relatives. Atta's English was far superior but he was little inclined to talk. He seemed to some people who met him to be what he sometimes said he was—a middle manager at a computer company. They tended toward the casual uniform of the Florida business class—polo shirts and khaki slacks—although Shehhi often looked unpressed and untucked. The two of them worked straight through the class regimen, including the extra training at nearby flight centers. In December, they were awarded license certifications to pilot small, single-engine, commercial aircraft.[35] That would qualify them for further training to fly jets. The training, the necessary first step on Khalid Sheik Mohammed's reconfigured plot, cost less than $30,000.

Unlike Atta and Shehhi, Jarrah made friends among his classmates. He even flew off to the Bahamas for a weekend with some of them. He told Aysel the friend who was the pilot on the trip drank too much and was incapable of getting them home. Jarrah, unlicensed and with just two months of classes taken, flew them safely back to Florida. "He was a friend to all of us," said Arne Kruithof, president of the flight school.

Jarrah bought a car—two of them, actually. The first was a lemon that he sold back to the dealer at a $1,000 loss, replacing it for $1,500 with a red 1990 Mitsubishi Eclipse.[36] Jarrah eventually moved in to an apartment with two of his classmates, sleeping on their couch because, he said, he had no furniture in his own apartment. He was typically the first to wake in the mornings and made tea for the house. He cooked for them at night, too. The other students said Jarrah, in spite of these indications of friendship, remained withdrawn, sometimes almost hostile. They said he had a temper, which could be explosive, and an absolute certitude about things he had no reason to be certain of—his flying abilities, for example. From very early in his lessons he was cocksure in the cockpit. One fellow student even refused to fly with him, fearing he wasn't careful enough.[37]

In October, Jarrah made his first visit back to Europe. He flew into Frankfurt, picked up Aysel, and the two of them went on to Paris, where they spent a weekend with Jarrah's sister, who lived there. They ate and drank and took in the sights.[38] Jarrah's family continued to finance his education, sending him $2,000 a month, and more if he asked. "I know that he wasn't working in Venice, but still he was living a high roller's life," Aysel said. "Whenever he or I needed money, I just had to call Ziad's parents, just tell them how much. Whenever I called up his parents and they asked how much I needed, they would always send over two or three times as much."[39]

During the next year, Aysel and Jarrah replayed the difficulties of their relationship again and again. Aysel was inquisitive. Jarrah was eva-sive. She couldn't find him. He didn't answer either of his two cell phones. They fought and made up. He came to visit. He left. When he returned to Florida after that Paris trip, Aysel heard nothing for days.

> Please, I ask you, please call me. Just give me a short call so I know that you're all right. I'm angry that you don't think about me and that I wait for a message here and have to think about you all the time. Can you think about me once and try to pre-tend to be me. You're taking so many risks and I know a lot even though you don't tell it. It's no surprise that I'm afraid for you, right?
>   I love you.
>   Your Aysel.

Ziad finally responded:

> I arrived well. I'm sorry I haven't sent you a message for a long time. I did get your letter and I found it super sweet. And full of understanding and compassion. It's not about trust. I love you, Aysel, and don't worry.[40]

In total, he returned to Germany five times in fourteen months while he was in the United States, and Aysel visited him once in Florida. They flew down to the Keys and he showed her how he trained in a Boeing simulator. He told her not to tell friends where he was. She agreed, but later said, "Of course I never stick to that. If somebody called and asked his whereabouts, I give them an answer. I told him, too, and he is very angry."

In December, Atta and Shehhi began checking out small aircraft, often at night and not returning until 2:00 or 3:00 A.M.[41] They took extra instruction on commercial flight simulators, which allow someone with rudimentary flying skills, such as they had acquired in their formal train-ing, to begin learning more advanced techniques without the expense and risk of flying a large jet. They took their first simulator lessons just outside Miami, in Opa-Locka. The six hours of lessons on a Boeing 727 flight simulator cost $1,500, which they paid in cash.

The day after Christmas, Jarrah went home to Beirut to visit his family. Early in the New Year, Atta and Shehhi left Florida, too. Atta flew out of Tampa for Madrid. The same week, Shehhi flew to Casablanca, Morocco. Jarrah spent a week in Lebanon, then came back to Germany at about the time that Atta was arriving via Madrid. Omar was just back from London. At least the three of them, and possibly Shehhi, too, were in Germany together for the last time.

HAMBURG

The men who had been left behind in Hamburg were not idle while their friends were training in the United States. After spending the first quarter of the year in Afghanistan, Omar had returned to Germany planning to join them in Florida. Just before he left Afghanistan, he had made a videotape record of his intent to become a martyr.[42] Before that could occur, however, he went searching for a bride. In early July, in Berlin, he met a young woman who, whatever else she might have been, was a very unlikely Muslim wife. She was a modern dance student, newly arrived in Germany and searching for a suitable school for dance training. She told Omar she was Roman Catholic and, although it is unclear, was apparently Japanese. She had come to Berlin because she had seen a German dance troupe on an international tour and been impressed.[43]

She met Omar on a Sunday afternoon in the lobby of the youth hostel where she was staying. She was reading notices on a bulletin board when she noticed Omar watching her. He introduced himself as Ramzi and asked her to go swimming, which she declined. But she accepted a lunch invitation and he took her by bus to an Arabic restaurant in downtown Berlin.

During lunch, he suggested a movie and they went to see the Russell Crowe film *Gladiator*. Afterward, when she complained about the cost of her stay at the hostel, where her bed was one of ten in a room, Omar said she could come with him to the student apartment where he was staying. She declined, in part because she had already paid for that night, but they agreed to meet the next morning.

Omar showed up wearing a suit and, after she checked out of the hostel, he took her by taxi to the apartment house where he had borrowed a room from a friend. There was not, as he had promised, a separate room for the woman, just one room with one bed for the both of them. The first night, Omar slept on the floor. The woman awoke the next morning startled to find him kneeling on the floor, praying loudly while grasping a huge Yemeni dagger.

They spent the next five days together. At the end of the first, Shibh proposed marrying the woman and taking her back to Hamburg. He was a university student there, he said. He said he had already told friends he was going to bring a wife home with him. She would, of course, have to change: to dress more modestly and cover herself in the way of a good Muslim woman.

Omar had previously grown angry when the woman had disagreed with him, so she didn't tell him outright that she would not marry him. She merely said she had come to Germany to dance and would only move to Hamburg if he found a suitable program there for her study. The temper, she said later, was most evident when he talked about non-Arabs. He refused to pay train fares because the Germans didn't deserve it, he told her; they were all racists. The Americans were as bad; he condemned them for their support of Israel. The Jews, he told her, were the worst.

"His great-grandparents, his grandparents, his parents hated the Jews and if he should have children, they would hate them, too," he said.

The woman said she told him the hating would have to stop at some point and he grew so angry she was afraid he was about to hit her. He screamed at her that his children must hate the Jews. For the most part, however, his temper tantrums were not the norm. He could be warm and funny and he told her she was the funniest person he had ever met.

Shibh's diet seemed to consist almost entirely of frozen pizza with tuna, which he bought at the supermarket and reheated at the apartment. During the four days, they roamed around Berlin. She window-shopped, he looked for telephone call centers and Internet cafés. He was constantly taking calls on his cell phone, then hustling to find a public phone to return the call, she said. He didn't explain what the calls were about.

They talked a lot about the movie they had seen and he later signed e-mails to her as, "Your King, Ramzi," a reference to it, she said. After five days, he returned to Hamburg and they never saw each other again.

He made three more attempts to obtain a U.S. visa. At one point, he wired a $2,200 deposit to FFTC, Jarrah's flight school, as a down payment on a similar training course. He used that application as a basis for a new attempt to get a student visa, rather than the visitor visa he had sought previously. He was growing desperate to join the others. On September 7, 2000, he sent a fax to FFTC, worriedly inquiring about the visa. In hand-printed block letters, he labeled the fax:

TOP URGENT

I wish to inform me by which company did you send me my visa document and which code nr has the mail got.

Please inform me today. It's too important. I'm waiting you for.[44]

For another attempt, he arranged for several thousand dollars to be deposited in his Yemen bank account, apparently to demonstrate financial wherewithal. This, too, failed. Consular officers at the American embassies in Berlin and Sana'a, where Omar made his applications, interviewed him at least three times, but each time determined he was a probable candidate to attempt illegal immigration. After the last failure, a consular official at the American embassy in Berlin advised Omar to quit trying: "Please acknowledge that we can not give you a visa."[45] The denials pointed out a characteristic failing of the U.S. border control regimen: It was focused overwhelmingly on preventing economic migration. People who were suspected of planning to take American jobs were barred; those, like Mihdhar and Hazmi, who were suspected of belonging to known terrorist organizations breezed through because the CIA had not thought to mention these suspicions to consular authorities.

Motassadeq and Essabar had by then returned from Afghanistan. Although they trained separately, the two Moroccans saw one another frequently at the camp near Kandahar. Essabar reveled in camp life; he spent nine months there and told Motassadeq he never wanted to go back to Germany.[46] He did return, however, in October, apparently called home by Omar. He immediately went through the same routine the others had upon leaving the camps. He applied for a new passport, claiming his had been lost. After receiving the new document, and with Omar failing yet again to receive visa approval, Essabar applied for a U.S. visa. The timing suggests the men were intent on finding a fourth pilot. Essabar, too, was rejected, likely for the same reason.

Omar after his final denial settled into a new role as the key contact between the would-be pilots—the hit teams, as they called them—and the main planner, Khalid Sheikh Mohammed, and the rest of the Al Qaeda hierarchy. Money passed to and from Omar. He and others in Germany took care to keep the men registered for university. It was not uncommon in German universities to be registered as a student and not take classes. University careers seemed at times to be exactly that—lifetime careers— so the prolonged absences seldom aroused suspicion or even notice. The men still in Hamburg even paid insurance premiums for the pilot trainees to make it seem as though they were still in the country. They handled their mail and whatever else arose.

In December, Omar flew to London. By this time, he had all but exhausted his efforts to obtain a U.S. visa. Essabar had been similarly unsuccessful. In London, Omar met with another pilot candidate—a French citizen of Moroccan descent named Zacarias Moussaoui.[47] As a French citizen, Moussaoui could come and go to the United States almost at will. Aside from that, he had more than a few problems. He was, most problematically, an erratic personality who, because of his public outspokenness about Islam, had already attracted attention from security services in France and Britain; it was likely the Moroccan service and one or two others had taken a look at him as well. Moussaoui, raised in France by a doting and upwardly mobile mother who rejected fundamentalist Islam,[48] had attended university in London and found a home in the city's flourishing radical Islamist scene. He went off to join the jihad, as a noncombatant, in Chechnya, then in a reversal of the usual path, went to Al Qaeda training camps in Afghanistan in 1998. After the camps, he returned to London as a crusading and very public advocate of radical Islam.

Whether in Afghanistan or after, Moussaoui—or more likely his

European passport—came to the attention of Al Qaeda's recruiters.[49] He was instructed to seek flight training and sent to Malaysia with the money needed to pursue it. He was hosted there by the same man at the same suburban Kuala Lumpur condominium who had previously helped Mihdhar and Hazmi, Hambali lieutenant Yazid Sufaat. Moussaoui had been told he could train at a flight school in Malacca, on the western coast. However, after a brief survey of the training opportunities, Moussaoui decided it rained too much and it would take too long to get what he wanted.[50] He returned to London. Before he left, Sufaat provided him with letters of reference attesting to his employment as a sales representative of Infocus, his wife's computer services firm. Moussaoui in September began e-mailing flight schools in the United States, including the Airman Flight School in Norman, Oklahoma, the school that Atta and Shehhi had visited that summer. On October 22, 2000, he e-mailed Brenda Keene, the admissions director:

> Hello, Mrs. Brenda,
>   I was very busy latelly [sic] to prepare my departure to the U.S. I hope to come in the next mont [sic] (maybe two weeks) if everything goes well. I think that when we meet (hopefully) we will have the opportunity to discuss further the financial situation, so for the moment I leave this. . . .
>   Fly well, bye, bye.

Rather than use his name, Moussaoui signed off the e-mail as he had all the previous ones: "zuluman tangotango."

Moussaoui met in London with Omar, who gave him instructions to go to Pakistan and meet with Khalid Sheikh Mohammed. Omar returned to Germany, and Moussaoui flew to Karachi, where Mohammed approved his proposal to go to Oklahoma.[51]

After Omar returned to Germany he met with Atta and Jarrah, although it is unclear if these were separate meetings or together. Shehhi and Omar are not known to have met. Shehhi had flown to Morocco and, like the others, could easily have gone on to Germany, too, but there is no record of it. The three pilots had completed their basic training, but not knowing what might happen with Moussaoui, they still lacked the fourth pilot Mohammed's plot envisioned. Essabar tried once more for a visa, but that was again unsuccessful. Later in the month, after the pilots returned to the United States, Omar flew to Tehran and from there made his way into Afghanistan.[52]

Mohammed didn't have high hopes for Moussaoui. He and Omar agreed that they would use him only as a last resort.[53] In any event, Mohammed had already found a candidate to fill Omar's pilot slot—a young Saudi named Hani Hanjour, who was already trained and licensed in the United States as a commercial pilot. Hanjour had been unable to find work as a pilot anywhere and had gone, instead, to the Al Qaeda camps in Afghanistan where word of his training filtered up in the organization.[54] In addition to his pilot training, his Saudi citizenship ensured he would have little difficulty getting into the United States. Mohammed quickly dispatched him to the States.

PHOENIX

Hanjour arrived in California in early December 2000. He was traveling on a student visa that was granted so he could study English at a school in Oakland. He never showed up at the school. He went instead to southern California, to San Diego, where Nawaf al-Hazmi had been living since his arrival from Malaysia in January.

Hanjour had lived in the United States off and on throughout the 1990s, mostly in Arizona, intermittently taking flying lessons at several different flight schools. He had first come to the United States, where his older brother lived, in 1990 to study English in Tucson. He stayed for three months, then returned to Saudi Arabia. The Hanjour family were prosperous merchants from Taif, a resort city in the Asir Mountains southeast of Mecca. Once the summer capital for the Saudi monarchy, Taif is now a vacation destination of choice for the children of the Saudi population explosion of the past three decades, during which the number of Saudi citizens has tripled. The outskirts of the city, which overlook a dramatic, rocky valley, are full of small amusement parks, Go-Kart tracks, even a monkey colony. The Hanjour home is in a pleasant, middle-class neighborhood in the hills above the city.

Hanjour returned to the United States in 1996. He had planned on becoming a flight attendant, but his brother, Abdulrahman Hanjour, told him that if he wanted to fly, he ought to become a pilot. He encouraged him to come to the United States and train. He arranged for Hani to stay initially with his friends Susan and Adnan Khalil. Abdulrahman was a lighthearted, sociable, easygoing man. Hani was his opposite—meek and quiet. He had no visitors. He prayed five times a day, and the only person the Khalils ever saw him with was another man who would take him to the mosque. During his stay, he barely uttered a word to the Khalils. "He was like a little mouse around the house," said Susan Khalil. "He really didn't make any friends, other than people giving him a ride to the mosque or from the mosque. When I was here, when my husband was at work, he would pretty much stay holed up in his room, probably reading the Qur'an or whatever. . . . He wanted to go back and get a job as a pilot with Saudi Airlines, or something like that. I remember him being excited about it. And I thought it would be great. I thought it would be good for him."[55]

Hanjour stayed with the Khalils for just a month, moved to California for language classes, then on to Arizona, where he enrolled at Cockpit Resource Management Training Center in Scottsdale, a Phoenix suburb. Duncan Hastie, the owner of CRM, said Hanjour attended the school the last three months of 1996. That began a sputtering progress toward a pilot's license. It took three years. "One of the first accomplishments of someone in flight school is to fly a plane without an instructor," said Duncan Hastie, CRM's owner. "It is a confidence-building procedure. He managed to do that. That is like being able to pull a car out and drive down the street. It is not driving on the freeway."[56]

Hastie described Hanjour as intelligent, friendly, and "very courteous, very formal," a nice enough fellow but a terrible pilot. Normally, a student could earn a license to pilot small, single-engine planes in three months, but Hanjour returned to Saudi Arabia that winter without a license. In the interim he called occasionally, Hastie said. "He was a pain in the rear. We didn't want him back at our school because he was not serious about becoming a good pilot." But he did come back the next year, failed again to get his license, then changed flight schools. He kept at it, even after being advised by a Federal Aviation Administration examiner that he would likely never get a commercial license because his English was so poor.

Abdulah Suliman, who roomed with Hanjour in 1998, said he was completely dedicated to his training. He studied, prayed, and went to the bank to get more money to pay for more training. "I'd tell him, 'Let's go see a movie.' He'd say, 'No.' I'd tell him, 'Let's go play basketball.' He'd say, 'No,'" Suliman said. "He just stayed home and studied his books for flight school."[57]

His perseverance was finally rewarded in 1999. After training in at least four schools, the FAA finally qualified him, granting him a commercial license. He returned home. As hard as he worked to get the license, he had an equally difficult time using it. He applied at several regional Middle East carriers without success. The failure depressed him. For much of the next year he did little but surf the Internet at a café his family owned, reading about flying. He had always been especially devout, and now he began listening to the cassette tape sermons of fiery imams that are ubiquitous in the kingdom.[58] Eventually, his listlessness was resolved in the direction of Afghanistan, and, like so many other young men, he took off for the camps. He arrived just as Khalid Sheikh Mohammed's men were looking for another pilot. They found Hanjour and he returned to the United States with an opportunity, at last, to fly.

Hanjour and Hazmi didn't stay long in San Diego after Hanjour's arrival. They left before Christmas. Hazmi told his landlord, Abdussattar Shaikh, he was moving to San Jose for school. They swung by the Texaco station in Hazmi's blue Toyota Corolla to say good-bye on the way out of town. In January, Hazmi called Shaikh, saying he had ended up in Arizona instead. "That's the last time I heard from him," Shaikh said.[59]

In Phoenix, Hanjour began flight training once again.

BALUCHISTAN

The desert camp was up out of the Kech Valley in the rocky foothills above Turbat, which is to say, it was very near to the middle of nowhere, on the leeward side of the Makran coastal range in the desert of far southwestern Baluchistan, about 75 miles from the Iranian border. Turbat itself, hardly a cosmopolis, was nonetheless home of one of the larger palaces of the Nawabs, the regional princes who rule much of Baluchistan, and was a transshipment point for Baluch dates, Iranian oil, and Afghan heroin and hashish. Up in the hills, however, the only sign of commerce was a camel-trading yard on the edge of the small desert village where the camp was located.

The men who were brought to the camp were a small and, by the time they arrived, rattled group of Pakistanis and Arabs. Most were simple, devout men who had thought they were making a sort of pilgrimage for religious training. They had been recruited out of the Gulf emirates by Tabligh missionaries, who promised to take them to revival meetings held annually in Pakistan. Instead, they ended up in this desert encampment after a roundabout journey that included a firefight in Kashmir and a three-day train-and-truck trek across the entire length of Pakistan. This was both much more and much less than they had bargained for.

Now that they were here, they were boarded in small mud huts in groups of three. There were about a dozen such structures in the camp. The last house was reserved for guests and the one in the center for meetings. They passed mainly pleasant days at the camp, praying, talking, listening, occasionally going into the village. At the end of the first week, after *isha*, or evening prayer, a half-dozen heavily armed men arrived in a small caravan of charcoal gray Toyota four-wheel-drive trucks. The leader was a portly, heavily bearded man who joined them in the center house and preached to them. This was no ordinary preaching and no ordinary preacher. He was calling the visitors to war, and his name was Khalid Sheikh Mohammed.[60] The Tabligh, a fast-growing Islamic revivalist sect, was founded in India in the 1920s. Its leaders stressed its apolitical, peaceful nature, but over the last twenty years, Al Qaeda has used the group's pretenses for its own recruitment.[61] Mohammed preached after prayers almost every evening for a month, beginning in mid-October, before the first snows fell. Said one of the men in the camp:

"He talks about how to live, what's going on in the world. Mostly he says he hates America, and the Jewish. He is very angry with the interference of America in every place—Afghanistan, Chechnya, Kashmir. 'That is too much, go kill the people.' That's what they said all the time. 'Go kill the people.'"[62]

Mohammed urged the Tabligh recruits to join his cause. He could get them new identities in Karachi, in minutes, he said, teach them how to change their appearances. The people with him talked constantly on Sanyo walkie-talkies. Some nights, men would come down from the mountains on horseback to meet with them. The leader of the mountain men, named Aref Sheikh, was always greeted warmly by Mohammed. Aref was taller but otherwise looked much like Khalid, enough so that people thought they were related. They probably were; the man could well have been Khalid's older brother, Aref, who was thought to have resettled in Baluchistan after the Afghan jihad.[63] The camp was just 50 miles or so south of the Mohammed family's ancestral home near Panjgur. Khalid Sheikh bartered with locals for shipments of the Iranian oil, siphoned away by the barrelful from storage tanks in Turbat. They talked about going into Afghanistan through Iran, avoiding Peshawar because the old routes were being closely watched. Khalid Sheikh was steadfast in meeting with the younger men every night to urge them on.

The year 2000 was a time of intense activity among the Al Qaeda elite, in particular for Khalid Sheikh. The revamped plan to attack the United States from the air had been set in motion. Excepting the difficulty in finding a fourth pilot, the early phases had gone extraordinarily well. Three pilot candidates had been found, financed, and sent to the States, and all had completed the primary portion of training without a hitch. A fourth pilot candidate was also found and dispatched. An effective communications regimen and hierarchy had been established, with Omar playing the role of intermediary. He was ideal for the job. Networking was what he did best and he had the confidence of both sides. He had known the pilot candidates for years. People who saw him in Afghanistan said a special bond seemed to have developed there between him and Bin Laden. The two of them met almost daily, one man said, and would often be seen engaged in intense conversation.[64] Bin Laden had demonstrated a tendency to employ Yemenis in his plots in numbers disproportionate to the numbers of Yemenis in his organization. He seemed to trust them implicitly.

Mohammed had other concerns, however. Al Qaeda had developed a very painstaking style of operation. Plots were allowed to take years to develop. For example, the first considerations for the East African embassy bombings began in the very early 1990s. People were actually put in place six and seven years before the 1998 attacks. The long lead times seem odd for several reasons, not least because they extended the security vulnerability over such a long period. Given that counterintelligence work seems to depend so much on good fortune, the longer an operation took, the longer the time luck had to interfere with it. This seemed not to have been a concern, or at least not one that had any effect.

A practical effect of the long lead times was that planning for several different operations occurred simultaneously. And some of the personnel in them overlapped. While the airline plot was gaining footing with the recruitment and dispatch of the pilots, Khalid Sheikh Mohammed spent three weeks in Rome, earlier in the year, presumably working on yet another plot.[65] He also assisted preparations for the attack on U.S. naval forces in the Gulf of Aden, in southwestern Yemen, led on the scene by Abd al Rahim al Nashiri and Tawfiq bin Attash—Khallad, as he was known.

Khallad had been a central figure at the meeting early in the year in Malaysia, the meeting that was spied on by the CIA. He later helped put together a new plan to attack an American ship. At least one of the men who had been with him at the Kuala Lumpur meeting, Khalid al-Mihdhar, was back in Yemen with him in the fall. Omar, too, was in Yemen, arriving the day before the Cole assault, leaving four days after.[66] This time, the bombers succeeded, running the same skiff, properly loaded this time, into the side of the USS Cole while it was anchored in Aden harbor. The explosion blew a huge hole in the side of the destroyer, very nearly sinking it. It killed seventeen sailors. Khallad and Mihdhar escaped without notice.

After the 1998 embassy bombings, the United States had responded by firing several dozen cruise missiles at a training camp in Afghanistan and a chemical factory in Sudan. Al Qaeda braced for the response this time. Bin Laden moved constantly throughout Afghanistan for weeks, often changing residences daily.[67] It was apparently why Mohammed was hiding out in Baluchistan.

The fact that Khallad was brought in to help direct the *Cole* attack, and that Mihdhar was involved, illustrates one notable aspect of Al Qaeda—how small the organization actually was. Over twenty-plus years, tens of thousands of men went through the Afghan training camps. In the same period, nearly a dozen attacks attributed to Islamic fundamentalists occurred around the world. But most of those men and most of those attacks had little, other than overlapping intent, to do with Al Qaeda. Most were independent groups running independent, often local, operations. In the attacks that were instigated by Al Qaeda, the same handful of people were involved in virtually every one. Even foot soldiers were recycled to new operations. The organization was so small that almost everybody in it at one time or another had personal interactions with top leadership. Men who went to the Al Qaeda camps almost casually would end up meeting Bin Laden. Some men who trained at the camps but never joined the organization reported repeated meetings with Bin Laden, Khalid Sheikh Mohammed, the military commander Mohammed Atef, and the recruiter Abu Zubaydah. Shared values enabled the organization to amplify its power by aligning with similarly politicized fundamentalist groups, many of them completely autonomous, around the globe. This made the group at times seem ubiquitous, but in fact it was a few men persistently pursuing a few deadly enterprises.

VIRGINIA

On October 20, 2000, Mohammed Belfas, Atta's Hamburg mentor and Omar's sometime roommate, accompanied Agus Budiman, a young architecture student he had known for years, from Germany to the United States. Budiman—like Belfas, an Indonesian—had been visiting the United States for years. He had family in the Washington, D.C., suburbs and now intended to move permanently to the area. Belfas later said he was an ordinary tourist who simply wanted to see the United States.

Budiman moved in with his brother and took a night job as a driver for Take-Out Taxi, a restaurant delivery service. During his two-week stay, Belfas often accompanied Budiman to work and offered to help drive the delivery car if Budiman would help him get a U.S. driver's license.

Budiman told Belfas he didn't need an American license to drive the delivery route. Belfas insisted, saying he wanted the license as a souvenir.[68]

On November 4, Belfas and Budiman made the first of two trips to the Department of Motor Vehicles office in downtown Arlington, Virginia. On the first trip, Belfas received a Virginia identification card after he and Budiman affirmed that Belfas lived in Arlington. They went back two days later and Belfas applied for and received a Virginia driver's license, using the ID card as proof of residence. Within the week, Belfas returned to Germany.

There, he said, he just happened to run into his former boarder, Omar, on a train. He told him about his trip and how he'd gotten the driver's license.

CHAPTER 3

The Last Year

THE LAST TASKS

WHEN THE HAMBURG PILOTS returned to Florida in January their focus shifted from the general job of learning to fly airplanes to the specifics of preparing for the attacks. They had several tasks:

- *Get enough flight time and hours in commercial simulators that they could actually fly an airliner, however briefly.*
- *Study airline procedures to determine how best to board and take over an airliner in flight.*
- *Communicate this to the men who would be sent to help them.*
- *House and feed those men, and assemble them into teams.*
- *Schedule the attacks.*

Immediately upon their return, they started more flight training.[1] Jarrah practiced extensively on a simulator in Miami. He had not finished the required work the year before to get his commercial license, as Atta and Shehhi had done. Atta and Shehhi showed up in a suburb of Atlanta, Georgia, where they began renting aircraft to fly. They did this off and on for a month, staying at a local discount motel.[2] They also made a brief trip to Virginia Beach, Virginia, where they opened an account at a commercial mailbox company one day, withdrew $4,000 in cash from their bank account the next, then returned to Florida.³ Across the country in Phoenix, Arizona, Hani Hanjour enrolled at JetTech Flight School.⁴ Instructors there worried about his poor language skills. English is the international language of the air industry; you can't fly safely without it. Hanjour held a commercial license, but his English was so bad the staff wondered how he got it and whether it was legitimate. They called local FAA inspectors, who confirmed Hanjour's license had been issued by them. The FAA suggested that Hanjour just needed a language tutor and let it go at that. Hanjour stayed at JetTech for several months. He spent most of his time on a Boeing 737 simulator, although he never progressed much.

About this time, Osama Bin Laden was urging Khalid Sheikh Mohammed to speed up the preparations. He wanted to get on with the attacks.⁵ Atta refused. There were things that had to be done first, he said, and Khalid Sheikh Mohammed backed him up. Bin Laden repeated the demands for months, but Mohammed and Atta wouldn't go before they were ready.

In February, Jarrah's father had heart surgery. The family asked Ziad to come home to be with him. Jarrah delayed, but scheduled a trip back at the end of March. On his return trip, he stopped in Bochum to see Aysel. His father's ill health appeared to have shaken him. He said he wanted to recommit himself to the relationship.

"He was really moved, and said, he, Ziad, wants to have children soon, so his father could see them before he dies," Aysel said.⁶

Later, after Jarrah returned to the United States and still wouldn't set a date when his training would end, Aysel grew angry again. Jarrah, as always, had an excuse; Aysel, as always, accepted it. In part, their relationship survived on her capacity to believe Jarrah's lies, even those that seemed preposterous. Once, on a visit back to Bochum, Jarrah showed Aysel a picture of him on one of his trips in a commercial airliner. The picture was likely taken on one of the reconnaissance flights the pilots took to examine in-flight security arrangements. On all of these, they flew first class or business class. Aysel asked why he was sitting in business class. Jarrah told her the flight attendant made him sit up front. Because I am Lebanese, he said, and they wanted me where they could keep an eye on me. This seems close to what Aysel wanted too—Ziad in a place where she could keep an eye on him. But even when he was within sight—in the same room or the same bed—Aysel saw only so far. Jarrah made it hard to look too deeply, but Aysel seemed to blind herself, too, not wanting to see what was in front of her. She saw Jarrah descend almost every step of the way into the September 11 plot. Even today she can recount the steps and still believe she didn't know where they were going. Of course, Aysel didn't believe the evidence. She believed what lovers always believe—she believed in Jarrah.

She was hardly alone in this. So public were the beliefs of the men from Hamburg and their associates that the often-stated notion that they were a secret cell of Al Qaeda "sleeper agents" appears to be almost opposite the truth. It sometimes seems that everyone who knew the men from Hamburg, especially their families, could see that something had gone seriously amiss in their lives.

The pilots each dealt differently with their families. Jarrah's parents were the only ones who had even an approximately correct idea of what he was doing—living in Florida, going to flight school. They, of course, didn't know why, but they were far ahead of the other families in their knowledge. Atta's parents knew he was in the United States, too, but they thought he was pursuing a Ph.D. in urban planning. Shehhi's family appears to have known nothing.

After he left the Emirates in the spring of 2000, Shehhi had called home regularly for a while, then less often. Finally, he quit calling altogether. When he didn't call at all that December during Ramadan, a serious breach of familial habit and responsibility, his mother (his father had died several years before) grew alarmed and tried to find out what was going on. She called the UAE embassy in Bonn several times, pressing them to find out what they could. The embassy made inquiries to Technical University in Harburg, where he was supposed to be studying. The university said Shehhi hadn't been there for a year; they had informed him by letter that he was being removed from the school's registration rolls. The local police had no record of anything untoward, but when it wasn't immediately apparent where he was they opened a missing-person investigation, which turned up nothing. Finally, Shehhi's mother dispatched his half-brother, Mohammed, to Germany. Mohammed and an officer from the embassy spent several days in Bonn and Hamburg searching for Marwan.[7]

"He asked everywhere, at the mosques there, even at an Arabic football club. No success," said Mohammed Awady, one of Shehhi's classmates from Bonn.[8]

"We knew he was not going to school and the Germans never had this," said a UAE security official. "We were trying to get him back. We were trying to track him."[9]

Finally, in Harburg they talked to Mounir el-Motassadeq, who had power of attorney over Shehhi's accounts. Motassadeq told them Shehhi had gone to Chechnya or Afghanistan.[10] Mohammed returned to the Emirates and later in the month Shehhi, apparently alerted by someone, finally called home. He told his family he had been going through a tough time, but that things were improving. He told them that Motassadeq was wrong about Chechnya; he had simply moved to a different area of Hamburg, but was still studying. He could, he said, see a light at the end of the tunnel. Whether or not they believed him, they let the matter drop and Shehhi returned to his work.

This core group of men involved in the plot, from a distance, seem to have done an enormous amount of traveling. From that distance, their time in the United States seems almost a blur of movement. It wasn't really. The three Hamburg pilots were in the United States for approximately fifteen months. For the first half of that period, they did little more than go to classes a few hours a day. In the second half of the period, most days they didn't even do that. By and large, they had long periods when they did very little. They waited. They ate lunch at the few Middle Eastern restaurants—usually Lebanese—they could find.[11] They went to the mall.

They changed residences every month or so, usually taking small furnished rooms or efficiency apartments in low-rent buildings. The three pilots had little contact beyond routine commercial interactions with locals and other travelers. When asked to list his permanent address on one rental application, Shehhi wrote: "None. I'm wandering."[12]

They traveled some. Some of the travel was to study airport and airline security, to find holes in it that could be exploited when the time came.[13] Atta, in particular, from his base in Florida, went to Oklahoma, Nevada, New Jersey, Virginia, California, Georgia, Massachusetts, and Maine. He also flew twice to Europe. Shehhi went with Atta to Georgia and Virginia and made separate trips to Nevada and California. He also left the United States twice—the initial trip to Morocco and possibly Germany and a second trip in the spring to Cairo, where he met with Atta's father.[14] He was away from the United States for two weeks, so it seems likely he would also have gone home to the UAE to pacify his mother. Jarrah traveled within the United States to Georgia, Nevada, California, New Jersey, and Maryland. He was by far the most frequent international traveler, with one trip to the Bahamas and five to Europe, one of those including an additional flight to Beirut.

Out west, Nawaf al-Hazmi stayed more or less at home in San Diego, with only occasional drives up to Las Vegas, Nevada. This changed when Hanjour arrived. They moved to Arizona for several months, then apparently drove across the country to Virginia, getting a speeding ticket in western Oklahoma en route. The Oklahoma traffic stop occurred, probably coincidentally, during the time another potential hijacker was briefly in flight school there—Zacarias Moussaoui.

Moussaoui, the erratic French Moroccan man Khalid Sheikh Mohammed had reluctantly approved for training, renewed his inquiries to the Airman Flight School in Norman, Oklahoma, in February, apologizing for the delay. "Life is never as you plan," he wrote. Moussaoui wanted to know the costs and length of training for a complete package, from ground school through commercial license. Then, all of a sudden, he e-mailed "Miss Brenda" at the school, telling her in late February: "URGENT. Flying to you tomorrow. . . . It will be nice if someone will be receiving me."[15]

Moussaoui arrived, rejected the proposed housing the school recommended as too public and too expensive, rented a room, bought a raggedy little car, and deposited $32,000 in cash in a local bank. He paid cash for a package plan aimed at getting a private pilot's license, a necessary first step. After three months and more than 50 hours of flight time, he still was not prepared to fly solo. The school informed him he'd be welcome to continue the course but he would have to start paying by the hour. He declined and said he would seek training elsewhere. If there was ever a plan to use Moussaoui in the September 11 plot, it likely would have ended right there.[16]

THE LAST MEN

B y late spring 2001, final preparations began in earnest with the arrival of the rest of the hijacking crews, the hit teams. These men, all but one of them from Saudi Arabia, started to arrive late in April. By the end of June they were all in the country. Or at least all of those who could get into the country had arrived. Khalid Sheikh Mohammed wanted even more men, as many as seven or eight per plane.[17] At least half a dozen men selected for the mission never made it into the United States—several had visas denied, others agreed to participate, then withdrew before ever leaving for the United States. At least one man was turned away by an immigration officer at arrival.[18]

Less is known about these late-arriving men, in part because Saudi Arabia has been parsimonious with the information in its hands, which is considerable, and has made the discovery of information by others difficult. But in a broad sense, it wasn't surprising that so many Saudis would be among the attackers: From the beginning, Saudis were the largest national group among the Afghan Arabs. Bin Laden is Saudi and so were many of the financial backers of the mujahideen and, later, the Taliban. The relief groups and charities that have been among the most prominent supporters of the Taliban and have been implicated in various Al Qaeda plots either were based in Saudi Arabia or derived much of their support from there.

Most of the Saudi hijackers were from the southwestern mountain province of Asir, a rugged, isolated area that is both physically remote and socially removed from the Saudi power structure. Saudi power is concentrated in the Jeddah-Mecca-Riyadh corridor. The rest of the country is effectively shut out, even the eastern portions that contain the overwhelming portion of the nation's oil reserves. Bin Laden, despite his family's great wealth, was made an outsider, too, by not having been from the royal tribes. He's Hadrami, people will say, meaning his family is from central Yemen, intending this as a slight and implying that's all you really need to know.

The southwest was entirely independent of Saudi control, with its own local dynasty, until the 1920s. Even after it became a part of the Saudi nation, it was never fully integrated into Saudi society and its people have never been fully assimilated. It has, however, received the fullest attention from the capital in religious matters. In the nineteenth century, missionaries and teachers swarmed through the region. They met great success. The southwest became and remains a stronghold of the strictest forms of Wahhabi Salafism, one of the most deeply conservative religious regions of the country. In that sense, its values are a heightened version of those typical throughout the kingdom.

Most of the young Saudi hijackers were from families headed by tradesmen and civil servants, well-off, but not wealthy. Several were described as among the best boys—bright, respectful—in their towns. They were largely unexceptional men and none of them stood out for their religious or political activism. Many had gone to university in Riyadh or Jeddah. Three had studied Islamic law. At least one, Ahmed Ibrahim al-Haznawi, just 21 years old, had already memorized the Qur'an, a sign of deep devotion much respected by others. One man, Wail al-Shehri, was a physical education teacher. Shehri had grown depressed, his family said, in late 1999. His father sent him to a local imam for advice. The imam prescribed recitation of Qur'anic verses to treat the depression. Not long after, Wail and his brother, Waleed, left home for Afghanistan.

Several other of the men left, like the Shehris, with close friends or relatives. Two-thirds of them told their families they were leaving to join the jihad. Several said they would fight in Chechnya. Others simply disappeared.

That young men from good backgrounds would leave homes and families without fanfare or discouragement was evidence of the broad support within Saudi Arabia for jihad. The flow of men that built up during the Afghan war against the Soviets had never stopped. With or without obvious enemies of Islam, the habit had been formed and, absent active interdiction by the government, continued without pause a decade after the original cause disappeared. The ordinariness of young men going off to war was also evidence of another fact—there was little or nothing for them to do at home. The Saudi Arabian economy had simply not produced worthwhile jobs for the growing ranks of Saudi young adults. With nothing to bind them, they were set adrift. With nowhere to go economically, they went where they were called by their religion.

Home-grown recruitment networks were embedded in the social structure of the country. They were an integral part of the society, not something overlaid on it. Saudi Arabia, as an especially powerful absolute monarchy, does not have politics—not in the sense that people in the West would understand it; there are no overt public debates of policy and national direction. But Saudi Arabia is also one of the most profoundly religious societies on Earth, and its religion is Islam, in which the religious and the political are inextricably bound. Everything political is religious; everything religious is political. And religion is much discussed. Waheed Hamza Hashim, a Saudi political scientist, said this created an organic recruitment process:

"In each street there is a mosque. In each mosque there is an imam. After evening prayer, every day, not every one stays. There is discussion. There is direction. There is a call. In Islam, everything has to be checked all the time. Is it right or is it wrong? Who do you check with? These sheikhs, some of these imams, are linked. . . . That's where it starts. Eventually you are transformed from a human being to a human bomb, a killing machine."[19]

There was no direct route for most men to go from the Middle East to Chechnya or Bosnia to fight, nor any real desire of warring Muslims in those regions to receive callow, untrained men. Those who left to fight holy wars, no matter where they imagined those wars to be, usually ended up in Afghanistan. None of this group of young Saudis ever made it to a traditional battlefield. They were selected by Al Qaeda recruiters on the watch for vigorous young men with clean passports and a willingness to die for their beliefs. The Martyrs Battalion, according to Khalid Sheikh Mohammed, was oversubscribed.[20] The original plan was to fill the ranks of the foot soldiers in the plot with men from a variety of countries, to apportion honor among them. Osama Bin Laden decided instead to send predominantly Saudis.[21] Whatever his reasons,[22] on a purely pragmatic level it was easier for Saudis than almost anyone else to get American visas. If this had ever been doubted, the experience of the hijackers proved it conclusively. Several of the young Saudis' passports had been tampered with in obvious ways, presumably to erase their travel to Pakistan. Most of them made substantial errors or omissions on their U.S. visa applications. Several failed to demonstrate the financial wherewithal for extended visits. None of these mistakes mattered.[23] They were all granted visas. The only non-Saudi was an Emirati compatriot of Shehhi's, Fayez Banihammad, who also had no problem gaining U.S. entry.

The men were trained in hand-to-hand combat in the Al Qaeda camps, taught the physical skills they would need for the sole task given them—to physically overpower flight crews. The pilots were the leaders. The new men would be the muscle.

Paradise was a powerful lure. Before they left Afghanistan, they made videotaped declarations of their willingness—no, eagerness—to become martyrs. Abdul Aziz al-Omari said in his tape that he relished what lay ahead: "I am writing this with my full conscience and I am writing this in expectation of the end, which is near. An end that is really a beginning. We will get you. We will humiliate you. We will never stop following you. . . . May God reward all those who trained me on this path and was behind this noble act and a special mention should be made of the Mujahid leader Sheikh Osama Bin Laden, may God protect him. May God accept our deeds."[24]

The thirteen men—the brigades of belief, Omar called them[25]—entered the fields of jihad in the most mundane way possible: through the cool plastic jetways and concrete garages of contemporary American airports. They came over a period of two months, flying to four different airports—suburban Washington, D.C., New York, Orlando, and Miami. They typically traveled in pairs. All came through Dubai in the UAE, where they were outfitted by Khalid Sheikh Mohammed's nephew with cash, credit cards, bank accounts, and some minimal idea of how to dress and behave in the U.S.[26]

The men who were already in the country—the four pilots and Hazmi—took the new men in hand upon arrival. Hazmi and Hanjour rented an apartment in Patterson, New Jersey, in late April. The four late arrivals who flew into Dulles and JFK were received by Hazmi there. The other nine, under the guidance of Atta and Shehhi settled in and around Fort Lauderdale, north of Miami. The three pilots from Hamburg moved across Florida from their apartments on the Gulf coast to the Atlantic to receive and guide them.

Aysel Sengün, typically, asked Jarrah to explain why he moved to the Miami area. He told her he had found a cheaper flight school, although in fact he was no longer even enrolled. He had left FFTC in Venice after getting his private pilot's license and an instrument rating, but he never completed the work for a commercial license. Aysel complained that Jarrah was hard enough to get hold of even when he wasn't moving. "Sometime I had to push him to give me a number," she said. Jarrah used prepaid calling cards for his phones, and more than once his time would expire in the midst of conversation with Aysel.[27] He didn't always call back.

From the arrival of the muscle on through early September, the hit teams lived on the suburban fringes of America, inhabiting a series of seedy apartments and motels. They bought memberships at gyms and worked out regularly. Jarrah even signed up for martial arts lessons. He took this seriously and could have been good at it, his instructor said.[28]

It is this period that gave the impression to investigators afterward of a sort of rushed, frantic movement. The groups rented cars, returned them, and rented others. They moved from one flyspecked town to another. But much of the movement seems to have been undertaken merely to avoid staying in one place so long they would attract attention. Other portions of it had very specific purposes. The pilots booked surveillance flights across the continent, examining operations and security on board different airlines. Hanjour and Hazmi went back to Arizona for a time so Hanjour could continue his work on the jet simulator.

The hijackers knew that local identification made the purchase of everything, notably airline tickets, easier to accomplish with less scrutiny. A number of them—seven of the nineteen—went to Virginia specifically to get identification cards or driver's licenses, having determined Virginia was a particularly easy place to do so. One way they might have known this was from Mohammed Belfas, the Hamburg man who had been a mentor to Atta. Belfas had gotten a Virginia license at precisely the same office in precisely the same manner the year before. Then, when he returned to Germany, he accidentally "bumped into Omar," Ramzi bin al-Shibh, and told him all about it.[29] Jarrah even used the same false address Belfas had used in applying for his identity card.

There were all sorts of retrospective sightings of the group or individuals within it. Atta, in particular, with his distinctive and later well-publicized face was seen everywhere. Or, at least that is what people remembered afterward. He was said to have inquired about buying a crop-dusting plane, to have been treated for anthrax poisoning, to have flown reconnaissance missions over nuclear power plants, to have turned into a two-fisted drinker at coastal taverns. Most notoriously, Atta was said to have flown under an alias to Prague, in the Czech Republic, in mid-April and to have met there with an Iraqi spy master plotting to blow up a broadcast tower belonging to Radio Free America. Little of this seems likely; some of it is demonstrably untrue, and none of it has been proved otherwise.[30]

THE LAST MEETING

By summer, almost everything was in place. On July 8, Atta flew to Spain to meet with Omar. Al Qaeda knew its electronic communications were susceptible to high-tech interception. If something important needed to be said, messages were passed by courier or in face-to-face meetings. One of the strengths of a small organization like Al Qaeda was just that—it was small. If Al Qaeda were a nation with all of the infrastructure that implies, it would have been more vulnerable to penetration by American intelligence. Its lack of power, in a way, made it more powerful. Its size also illustrates the nature of asymmetric threats posed by nonstates. The September 11 attacks were by far the biggest thing it had ever attempted, but even at that, the number of people involved in the plot could be counted by the handful. The scale helped keep it hidden.

Atta, by that point, must have been anxious. The day before he left Florida, he dialed a cell phone in Germany seventy-four times.[31]

Presumably, the phone belonged to Omar and Atta wanted to ensure their meeting. The day after Atta arrived in Madrid, he rented a silver Hyundai and set off for Tarragona, an eight-hour drive east to the Gold Coast, as the Spanish call that portion of the Mediterranean shore. The same afternoon, July 9, Omar boarded a vacation charter direct flight from Hamburg to Reus, a small airport near Tarragona. The Gold Coast is a sunny, down-scale tourist area, a poor cousin to the more glamorous resort areas farther up the coast. The winter population of Salou, where Atta booked a room, is 16,000 people, a third of them Moroccans. In the summer, the population swells to 300,000. Atta and Omar arrived in the midst of them, mostly tourists from England and the European interior.

After a night at hotels in separate adjoining towns, Atta and Omar disappeared for a week. Except for Atta making a couple of trips to ATMs and booking a return flight to Florida, there is no trace of them—no hotel bookings, restaurant charges, or telephone calls. The assumption is that they met at a safe house provided by the extensive Islamist network that had been constructed over the previous decade in Spain. The network was headed by Imad Eddin Barakat, another of the exiled Syrian Muslim Brotherhood members scattered across Europe.

Barakat was a big guy, beefy, and loud. He made a living as a peddler, at times on the street. He sold used cars, leather jackets, and knock-off Lacoste shirts out of the trunk of whatever car he was driving at the moment. He sold cassette tapes at subway stops. "I work everything: clothes, cars, honey, rugs," Barakat said. "Whatever turns up, I'm a free merchant, I don't have one thing."[32] Although he never seemed to make much money, he came up with plenty when he needed it and lived beyond his apparent means in a top-floor condominium in a solid working-class neighborhood. He somehow had the wherewithal to bankroll construction projects and purchase a retirement home for senior citizens in the mountains outside Madrid; the old-folks home was unusual in that it had a large Arab staff and not many old folks.[33]

A police investigator likened him to royalty among his group: "He's like a monarch who never carries money because others carry it for him."[34]

Like many leaders in Islamist networks, Barakat traveled widely and often: to London (more than twenty times in one year in the 1990s),

Turkey, Belgium, Italy, Jordan, Syria, the United Arab Emirates, and Malaysia, hardly the travel regimen of the humble peddler he made himself out to be.

As Spain had aged demographically and boomed economically in the 1990s, immigration shot up. Moroccans dominated a North African influx that created migratory and transport networks from the coasts where illegal entrants wash ashore in smugglers' rafts. The sudden Muslim population created a pool of potential recruits among restless young men scraping to get by in construction, agriculture, and service industries.

In 1994, Barakat founded the Soldiers of Allah, a fundamentalist group that clashed with the imam of the mid-city Abu Bakr Mosque in Madrid.[35] Police say Barakat's group and an allied North African network offered a support structure and safe houses spreading from Madrid to Pamplona and Granada. Barakat could hardly have been unaware of Atta and Omar's presence in his territory. He was good friends with two fellow Syrians in Hamburg, Mohammed Haydar Zammar, who is suspected of recruiting Atta and Omar for Al Qaeda, and Mamoun Darkazanli. Darkazanli, in fact, was in Spain at approximately the same time as Omar and Atta, and he and Barakat had made plans to rendezvous.[36]

Wherever it was that Atta and Omar met, the purpose of the meeting was mainly the delivery of a status report from Atta to be relayed from the teams in the United States to the command in Afghanistan. Atta told Omar the teams were prepared, the targets researched, and the pilots ready. They discussed the final preparations and the means by which Atta would notify Omar of the date of the attacks, which had yet to be set.[37]

There was also a problem. Ziad Jarrah was having difficulties. It is no longer possible to know precisely why, but the relationship between Jarrah and Atta deteriorated to the point that Jarrah threatened to withdraw from the plot.[38] The two had never known one another well. In Hamburg, Jarrah attended a different school and lived by himself in a different part of town. He was closer to Omar than to anyone else. Omar was supposed to train with Jarrah in Florida, but when he failed to gain entry Jarrah was left on his own, and he refused to submit to Atta's authority. While Atta and Shehhi plowed through their training, Jarrah for whatever reason stopped short of getting his commercial pilot's license. One obvious difficulty was Jarrah's continuing relationship with Aysel. This bothered the ever-disciplined Atta, Omar said.[39] Atta, remember, was bothered when members of his group in Hamburg rushed through their Qur'anic readings. Not doing things the way they ought to be done could not be tolerated. Atta reported the problems to Omar. Omar relayed them to Khalid Sheikh Mohammed, who was displeased. He ordered Omar to make peace between the two because they could not afford to lose Jarrah now.[40] In an e-mail, Mohammed warned that if Jarrah "asks for a divorce, it is going to cost a lot of money."[41] However he did it, Omar made peace and the march toward the end, staggering now at times, continued.

Shortly before returning to the United States, Atta sent a cell phone text message to Said Bahaji and two other friends in Hamburg: "Salam. This is for you, Abbas and Mounir. Hasn't the time come to fear God's word. Allah. I love you all. Amir."[42]

THE LAST MONTH

Zacarias Moussaoui, as usual, e-mailed ahead. This time, he contacted the Pan Am International Flight Academy, a chain of flight schools headquartered in Florida:

Hello, I am Mrs. Zacarias!

Basically, I need to know if you can help to achieve my 'Goal' my dream. I would like to fly in a "professional" like manners [sic] one of the big airliners. I have to made my mind which of the following: Boeing 747, 757, 767, or 777 and or Airbus 300 (it will depend on the cost and which one is the easiest to learn).

The level I would like to achieve is to be able to takeoff and land, to handle communication with [Air Traffic Control], to be able to successfully navigate from A to B (JFK to Heathrow for example).

In a sense, to be able to pilot one of the Big Bird, even if I am not a real professional pilot. . . . I have around 55 hrs of fly in a [Cessna] 152, and I passed my written [exam] last month.

I know I could be better but I am sure that you can do something. After all we are in AMERICA, and everything is possible. Have a nice day, waiting for a positive fly.

Thanks you [sic]

Zac[43]

On July 10, Moussaoui registered at the Pan Am International Flight Academy in suburban Minneapolis for a flight simulator course. He was still in Oklahoma, where he had dropped out of his earlier course, but he remained on Al Qaeda's list as a potential pilot. Omar wired him another $14,000 at the beginning of August. Moussaoui and a friend, Hussein al-Attas, drove to Minnesota. They didn't plan to stay long. Moussaoui told Attas they would have to leave the country for Pakistan by the second half of September at the latest. In the meantime, Moussaoui would do two things: learn to fly big jets and teach Attas to fight. You had to know fighting if you were going to join the jihad, Moussaoui said.[44]

They rented a small motel room and Moussaoui paid $8,300 to the flight school. On August 13, he started his new training. He didn't get far. Two days later, the school called the local FBI office, telling agents they had what they thought was a potential hijacker on their hands. He had some training on small, single-engine planes and now insisted on learning to fly—quickly—the biggest airplanes in the world's commercial fleets. By the next afternoon, the FBI was at the school. By nightfall, Moussaoui was in jail on immigration charges. By the following day, the Minnesota agents had alerted counterterrorism officials in Washington and began seeking permission to search his belongings, specifically his laptop. One agent even wrote in the margin of his interview notes that Moussaoui was the type of guy who might hijack an airplane and fly it into the World Trade Center. The concern was great enough, and the activity so far from routine, that both FBI director Louis Freeh and CIA director George Tenet were briefed on the arrest. Nonetheless, the Minnesota agents never got permission to search Moussaoui's computer. In it, there were records of his dealings with Omar and Hambali.

No one in the U.S. government at that point knew, of course, that Khalid Sheikh Mohammed had planned an attack on the United States and the attack was imminent. But the government did know somebody from Al Qaeda wanted to do something. Intelligence networks for months had been ablaze with ominous information, hints and suspicions. Throughout the intelligence communities, both in the United States and abroad, there was an air of acute concern, verging on panic in places, that something very bad was about to happen. The signs were there.[45] The world's intelligence machines produced prodigious amounts of information, so much so that analysts habitually struggled just to keep up with the flow much less take time to consider what was in it.[46] Electronic intercepts, telephone chatter, warnings from foreign services and informers, internal memos— everything pointed in one direction: there was something out there. Some of the information even concerned Khalid Sheikh Mohammed specifically. The CIA had received credible reports that Mohammed was recruiting agents to travel to the United States and, once there, establish contact with agents already in the country.[47] This was, of course, exactly what he was doing.

"As I recall, during the period January to September 2001, the FBI received over 1,000 threats," said Thomas Pickard, deputy director of the FBI.[48] "Many of these threats had great specificity and others were very general in nature. The increase in the chatter was by far the most serious, but it was also the most difficult to deal with."

"Our collection sources 'lit up' during this tense period," said CIA director George Tenet. "They indicated that multiple spectacular attacks were planned, and that some of these plots were in the final stages. . . . But the reporting was maddeningly short on actionable details. The most ominous reporting, hinting at something large, was also the most vague. . . . Our analysts worked to find linkages among the reports, as well as links to past terrorist threats and tactics. We considered whether Al Qaeda was feeding us this reporting—trying to create panic through disinformation— yet we concluded that the plots were real. When some reporting hinted that an attack had been delayed, we continued to stress that there were, indeed, multiple attacks planned and that several continued on track. And when we grew concerned that so much of the evidence pointed to attacks overseas, we noted that Bin Laden's principal ambition had long been to strike our homeland."[49]

All the while the community was being deluged with frightening information it could not decipher, FBI agents in Minnesota were desperately and unsuccessfully trying to get court approval to look at Zacarias Moussaoui's computer and address book.[50] Other agents had been belatedly notified by the CIA that two suspected Al Qaeda agents, Nawaf al-Hazmi and Khalid al-Mihdhar, might be in the United States. They began to search for them a full nineteen months after they first arrived.[51] CIA analysts had finally realized that the men Hazmi and Mihdhar had met with in Malaysia in January 2000 were involved in the bombing of the USS *Cole* the following December.

The two Saudis by that time had been fully integrated with the pilots from Hamburg and the thirteen young men who had arrived in the spring. The nineteen Arabs moved in varying combinations up and down the East Coast for several weeks. They worked out at gyms, ate cheap pizzas, prayed, and studied airline flight manuals.

Everything was set. Things had gone more or less according to plan, although they were still short one man. The hit crews were intended to number five per aircraft, twenty men in all. They only had nineteen. The intended twentieth man, a part of the late spring Saudi detachment, had been denied a visa in Saudi Arabia and subsequently never came to the States.[52] Khalid Sheikh Mohammed made at least a couple of attempts, and perhaps more, to replace him. In July, his nephew Ali Abdul Aziz Ali, who had handled much of the chore of processing wire transfers to the hijackers in the United States, applied for a U.S. visa in the Emirates. He was turned down by American consular officials, again not because of any terrorist affiliations but because he was suspected of being a potential economic migrant. Khalid Sheikh Mohammed himself, using a Saudi identity, applied for a U.S. visa.[53] He received it, but there is no evidence he attempted to use it.

As late as August, Mohammed al-Katani, another Saudi, made it all the way to Orlando before he was turned away at immigration. The man possessed what was deemed to be insufficient funds for the week-long vacation he said he intended to take, had no return ticket, no credit cards, and a very bad attitude. He told an INS inspector at the airport that a friend was waiting outside to pick him up; then, when pressed, he reversed himself and said there was no friend. Atta was upstairs in the terminal at the time, apparently waiting for Katani.[54] The man never made it out of the immigration inspection area. He was put on the first flight back to Dubai. With him went the final reason to wait.

They would proceed with the nineteen men on hand. They would take four airplanes to four targets, three of which had long been selected: the twin towers of New York's World Trade Center and the Pentagon. There was vigorous debate for months on whether the fourth and final target should be the White House or the U.S. Capitol. Bin Laden preferred the White House, while Atta argued it was too difficult a target to hit.[55] He eventually won out, and the final date was chosen to ensure that the members of Congress would be back in session, and, thus, targets themselves. In mid-August, Atta messaged Omar through an Internet chat room: "The first semester commences in three weeks. There are no changes. All is well."[56] The hit crews began reserving and purchasing air tickets. Omar waited for a confirming call from Atta to be sure. It came on August 29. Omar, in turn, sent Zakariya Essabar to Afghanistan to give Khalid Sheikh Mohammed the news.[57] Zero hour, as they called it, would be September 11, a Tuesday.

News went out over the Al Qaeda network. In Pakistan, Mohammed told a young recruit he had been prepping for a mission in Southeast Asia that he needed to be in place by September 11, implying it wouldn't be safe to travel after.[58] In Europe, the remaining members of the Omar-Amir group in Hamburg made exit preparations as well. Three months before September 11, Said Bahaji had told his employers at the computer company that he would be quitting his job in the fall. He had accepted an internship in Pakistan, he told them, and would be moving. His employers said he was an exceptional worker. They were sorry to see him go.[59]

Bahaji told his family the same story about the internship. His aunt, Barbara Arens, heard the plan and didn't believe a word of it. She went to the police before September 11 to try to get them to do something.[60] Like what, they asked. Bahaji left Hamburg on September 4, flew to Karachi via Istanbul, and disappeared. Two passengers traveling on the same flight stayed in the same room with Bahaji at the Embassy Hotel in Karachi. Both were Algerians. One of them had been living in the same Spanish town Atta and Omar traveled to in July.

Zakariya Essabar and Omar disappeared from Hamburg, too. Omar left for Spain on September 5, flying from Dusseldorf. He stayed at a private home in Madrid, obtained a set of false identity papers, then flew to Greece, to the UAE, to Egypt, then to Afghanistan and off the map.[61]

THE LAST NIGHT

Over the next week, the men who made it to America moved into position, taking rooms in motels and hotels in or around Washington, New Jersey, and Boston. Atta, inscrutable as ever, checked out of his Boston-area hotel with Abdul Aziz al-Omari on September 10, got in a rental car, and drove to Portland, Maine. The two of them spent the evening shopping at Wal-Mart and eating at a Pizza Hut before turning in. No one outside the plot knows why.[62]

Several of the Saudi men in Boston made a series of telephone calls trying to arrange for prostitutes on the last night. In the end, they thought the prices were too high and didn't employ anyone.

Ziad Jarrah, at a Day's Inn in Newark, wrote one last letter to Aysel. For once, he did not hide from her.

Hello my dear Aysel

My love, my life. My beloved lady, my heart. You are my life.

First of all, I want you to believe truly, and really take care that I love you from all my heart. You should not have any doubts about that. I love you and I will always love you, until eternity. I don't want you to get sad I live somewhere else where you can't see me and can't hear me, but I will see you and I will know how you are. And I will wait for you until you come to me. Everyone has his time and will go then. I am guilty of giving you hope about marriage, wedding, children and family. And many other things.

I am what you wish for, but it's sad you must wait until we come back together. I did not escape from you, but I did what I was supposed to. You should be very proud of me. It's an honor, and you will see the results, and everybody will be happy. I want you to remain very strong as I knew you, but whatever you do, head high, with a goal, never be without goal, always have a goal in front of you and always think, "what for."

Remember always who you are and what you are. Keep your head high. The victors never have their heads down!

Hold on to what you have until we see each other again. And then we will live a very nice and eternal life, where there are no problems, and no sorrow, in castles of gold and silver and, and, and . . .

I did not leave you alone. Allah is with you and with my parents. If you need anything ask him for what you need. He is listening and knows what is inside you.

Our prophet said: "He is a poor man, who has no wife and she is a poor woman, who has no man." I will pick you up anyhow and if you marry again, do not have fear. You know, I don't like all men. Think about what you are and who could deserve you.

I hug you and I kiss you on the hands. And I thank you and I say sorry for the very nice, tough five years, which you spent with me. Your patience has a price. . . . God willing, I am your prince and I will pick you up.

See you again!!

Your man always

Ziad Jarrah

9/11/2001[63]

Since September 11, there has been much speculation that many, perhaps most, of the hijackers did not know what was about to happen, did not know they were about to die. They knew. Distributed among them was a handwritten set of instructions, admonitions, suggestions, and encouragements. The notes calmly advised meticulous preparations in the way that the Prophet's companions prepared themselves for battle twelve centuries before. Omar said the notes were prepared by Abdul Aziz al-Omari, the Saudi who accompanied Atta to Maine.

"These words," Omar said, "do not come from someone who I suppose to be deceived with vain hopes, as the hypocrites and the people of hypocrisy claim. It comes from a man who knows what he is doing and what he is about to do; not an ignorant man, but a man who is a believer, where all that is hidden and all that is seen have become equal to him. Everything is the same."[64]

The heading on a portion of the notes was "The Last Night." It contained fifteen specific instructions. The first was "Vow to accept death." The last was to wash in the morning so that "the angels seek forgiveness for you." In between, the hijackers were admonished to know the plan, sharpen their knives and prepare for the ascent to paradise where they would be welcomed as God's eternal companions; "the time for playing has passed, and the time has arrived for the rendezvous with the eternal Truth."[65]

Indeed, the time had arrived.

CHAPTER 4

That Day

BEFORE

IT WAS AN EXCEPTIONALLY FINE DAY for flying: bright early autumn blue skies, sharp-angled morning light, visibility unlimited.

"These hours were awesome, for you engage in a great battle with all its dimensions, a huge battle," Omar said. "It is a military operation that is unconventional against the mightiest force on Earth, who possess all the weapons and intelligence equipment and the spy satellites while her agents are spread all over the world. And you are facing them on their own back-yard, amidst their forces and their soldiers, with a group of youths numbering nineteen."[1]

The group of youths numbering nineteen proceeded that morning to four battlefields:

American Airlines Flight 11, a Boeing 767 with 81 passengers and a crew of 11, out of Boston Logan bound for Los Angeles International, scheduled departure 7:45, actual departure 7:59.

United Airlines Flight 175, a Boeing 767 with 56 passengers and a crew of 9, out of Boston Logan bound for Los Angeles International, scheduled departure 7:58, actual departure 8:14.

American Airlines Flight 77, a Boeing 757 with 58 passengers and a crew of 6, out of Washington Dulles bound for Los Angeles International, scheduled departure 8:10, actual departure 8:20.

United Airlines Flight 93, a Boeing 757 with 38 passengers and a crew of 7, out of Newark International bound for San Francisco International, scheduled departure 8:01, actual departure 8:42.

Atta and Omari were up before dawn in Maine to catch a 6:00 A.M. commuter flight to Logan. Their motel was next to the airport, but still, they cut it very close. They left the Comfort Inn at 5:33, drove to Portland Jetport, left their rented Nissan Altima in the lot, and passed through security inside the terminal just seven minutes before takeoff.

That morning, also very early, Ziad Jarrah, near Newark International, called Aysel Sengün one final time. She had complained often that even when he did call, the conversations were brief, sometimes cut off when his prepaid calling cards ran out of time. This conversation was abrupt even by those standards. Three times, quickly, he told her he loved her. Then Ziad told Aysel good-bye.[2]

Jarrah also called Marwan al-Shehhi in Boston that morning.[3] Shortly after, both men headed for their final flights. Shehhi had spent the prior 24 hours tidying up. He and Atta had shared breakfast at the coffee shop of the Milner Hotel the previous day before Atta took off for Maine. Later in the afternoon, Shehhi wired $5,400 in leftover funds back to the United Arab Emirates. The next morning—the last day—he checked out of the Milner, a budget business hotel, and drove a white Mitsubishi Mirage compact to Logan Airport, where he and his passengers got into an argument over a parking spot in the underground garage.

Nawaf al-Hazmi, Khalid al-Mihdhar, and Hani Hanjour checked out of the Marriott Residence Inn in Herndon, Virginia, and drove the 1988 Toyota Corolla Hazmi had bought in San Diego to Dulles International. They left the car, which was littered with flight training manuals, cockpit diagrams, and maps, in an hourly parking lot and checked in at the American Airlines ticket counter.

The hijackers' first task that day was to get on the airplanes with their weapons at hand. U.S. airport physical security was an acknowledged sieve before September 11. Every test that had ever been done on it demonstrated the ease with which almost any sort of contraband material could be gotten on board. The morning of September 11 was no different in that regard than any other day. The nineteen men carried chemical sprays, utility knives, and box cutters. The utility knives were permissible under FAA regulations. The box cutters and sprays were not. They all made it on board.

"All nineteen hijackers were able to pass successfully through checkpoint screening to board their flights. They were nineteen for nineteen, 100 percent," said John Raidt, an investigator.[4]

Several of the hijackers set off alarms at security checkpoints. Some set off two sets of alarms, but were waved with hand wands and allowed to board.

American air carriers, in addition to the physical security checkpoints, attempted to use a computerized system to select certain passengers for extra preflight screening. Factors the computer considered in selecting passengers for more rigorous scrutiny included purchase of one-way tickets, cash purchases of tickets, and, most obviously, any indication from government sources that the passenger was dangerous and should be barred from flight. The Federal Aviation Administration had compiled what it called a no-fly list. None of the nineteen men was on it. This was not surprising. For four years after Khalid Sheikh Mohammed had been indicted for trying to blow up a dozen American airliners, his name was not on that list. Hardly anyone's was. That day, the FAA no-fly list had on it just twelve names. By comparison, the State Department maintained a watch list of people whose entry and exit into the country ought to be monitored and in many cases prohibited. That day, that list had 61,000 names on it. The airlines did not have access to that list.

American commercial airlines carry more than 2 million people a day. Delay is a cost that frequently is never recouped. Delay invariably means fewer seats are going to be filled by passengers. A seat that goes out unfilled on a commercial flight is said within the industry to have spoiled. It's no good and must be thrown away, and with it goes whatever money the airline might have earned had it been filled. With the aircraft, on the other hand, goes the cost of flying that empty seat, which happens to be almost identical to the cost of flying it filled. When the plane takes off, the airline has already spent the money to fly it whether the seat is full or not.

The point is: airport security, while theoretically intended to provide safe flight, was more realistically intended to provide safe flight so long as it did not unduly impede people getting on airplanes and thus cost airlines revenue. This is not an idle consideration. Every FAA rule must pass a cost-benefit analysis before being enacted. For many years, the FAA had employed more economists than security inspectors. The main reason the FAA did not—and does not—incorporate the State Department's watch list into its own security lists is the belief that the airlines would either be unwilling, because of cost, or unable, because of cost, to check it.[5]

That day, nine of the nineteen hijackers nonetheless were selected by the computer for secondary screening at airport security. For most, this was the result of having bought one-way tickets, having paid cash for their tickets, or having provided insufficient or inaccurate identification when the tickets were purchased. The consequence of being selected for screening was that their checked baggage would be X-rayed and would not be placed on the plane until all the passengers, including especially them, had boarded. The theory underlying the baggage screening was that no passenger on board a flight would detonate a bomb in his or her luggage since that passenger, too, would be at risk. Three of the selected hijackers had no checked baggage, so they did not even suffer the indignity of delay. The secondary screening for them had no consequence whatsoever. The FAA theory of protecting only against nonpassengers did not countenance passengers who, as Omar put it, were looking forward to "the scent of heaven beyond the battle."[6]

In any event, that day the FAA did not unduly impede any of the nineteen men. Many, if not most, of the hijackers carried knives of sufficient size and heft to kill a man. All of them passed through security.

The plan was to take over the planes as soon as possible, within fifteen minutes after takeoff, Omar said: "The work was divided inside the aircraft. A group was assigned the task of storming the cockpit and the other group's task was to support the first group and provide protection from the back and prevent any attempt at foiling the process of storming the cockpit. Obviously, the group storming the cockpit is formed of two persons. It would be the nearest group to the cockpit. In order to seize the opportunity when the door is opened and enter into it swiftly, take it over, and slaughter all those inside and then the brother pilot comes very quickly to assume the rest of the mission and guide the aircraft. . . . All these actions must take place in record time, six minutes at the most."[7]

The hijackers acted as soon as the seat belt signs were turned off and the cabin crews began beverage service. They used chemical sprays, utility knives, and, in just one known case, Flight 77, box cutters. On three of the flights, they also used the threat of a bomb. Again, only Flight 77 was the exception to this. On all four flights, the hijackers killed or disabled people who got in their way. They herded the rest to the rear of the aircraft. Word of this was relayed surprisingly quickly to the ground by alert cabin crews. On American 11, attendant Betty Ong called the airline reservations system to report her flight had been hijacked at 8:21, just nineteen minutes after it lifted off the runway. Other calls were made on both her airplane and the others.

Response on the ground was incredulous, confused, and unhurried. Never having been forced into a situation remotely like the one it faced that morning, the system failed to do anything useful. From a national defense standpoint, this is what happened, according to retired Air Force Colonel Alan Scott of the North American Air Defense Command, NORAD, as it is called:[8]

8:02 – American 11 took off from Boston.
8:16 – United 175 took off on the same approximate ground path.
8:19 – American 77 took off westbound out of Dulles.

"So now the first three airplanes are airborne together," Scott said. "The first time that anything untoward, and this was gleaned from FAA response, that anything out of the ordinary happened was at 8:20, when the electronic transponder in American Airlines 11 blinked off, if you will, just disappeared from the screen."

A transponder is an electronic device that emits a signal that can be read by ground radar, conveying who, what, and where the plane is, its heading and speed. Ground radar can track a plane without a transponder, but it becomes much more difficult and imprecise.

"At 8:40 in our logs is the first occasion where the FAA is reporting a possible hijacking of American Airlines Flight 11. And the initial response to us at that time was a possible hijacking had not been confirmed. At that same moment, the F-15 alert aircraft at Otis Air Force Base, Massachusetts, about 153 miles away, were placed immediately on battle stations by the Northeast Air Defense Sector commander.

"At 8:43, as this is going on, the fourth airplane, United 93, takes off out of Newark, New Jersey.

"At 8:46, the last data, near the Trade Center, 8:46, the first impact on the Trade Center. At that minute is when the Otis F-15s were scrambled. And, again, they were 153 miles away. . . . Those F-15s were airborne in six minutes.

"At 8:53, that's a minute later, in the radar reconstruction, we are now picking up the primary radar contacts off of the F-15s out of Otis.

"At 8:57, which is seven minutes after the first impact, is, according to our logs, when the FAA reports the first impact. And about this time is when CNN coverage to the general public is beginning to appear on the TV, not of the impact, but of the burning towers shortly thereafter. So you can see what in the military I am sure you have heard us talk to the fog and friction of war, and as the intensity increases the lag tends to also increase for how quickly information gets passed.

"9:02—United 175, the second airplane, which by the way never turned off its transponder before impact, crashes into the [South] Tower at 9:02. The distance of those fighters which had been scrambled out of Otis, at that particular point, they were still 71 miles away, about eight minutes out, and going very fast.

"At 9:05, FAA reports a possible hijack of United 175. Again, that's three minutes after the impact in the tower. That's how long it is taking now for the information to flow through the system to the command and control agencies and through the command and control agencies to the pilots in the cockpit.

"At 9:09, Langley F-16s are directed to battle stations, just based on the general situation and the breaking news, and the general developing feeling about what's going on. And at about that same time, kind of way out in the West, is when American 77, which in the meantime has turned off its transponder and turned left back toward Washington, appears back in radar coverage. And my understanding is the FAA controllers now are beginning to pick up primary skin paints on an airplane, and they don't know exactly whether that is 77, and they are asking a lot of people whether it is, including a C-130 that is westbound toward Ohio.

"At 9:11, FAA reports a crash into the South Tower. You can see now that lag time has increased from seven minutes from impact to report; now it's nine minutes from impact to report. You can only imagine what's going on on the floors of the control centers around the country.

"At 9:24 the FAA reports a possible hijack of 77. [The actual report was for American Airlines Flight 11.] That's sometime after they had been tracking this primary target. And at that moment, as well, is when the Langley F-16s were scrambled out of Langley.

"At 9:25, American 77 is reported headed towards Washington, D.C., not exactly precise information, just general information across the chat logs.

"9:27, Boston FAA reports a fifth aircraft missing, Delta Flight 89— and many people have never heard of Delta Flight 89. We call that the first red herring of the day, because there were a number of reported possible hijackings that unfolded over the hours immediately following the actual attacks. Delta 89 was not hijacked, enters the system, increases the fog and friction if you will, as we begin to look for that. But he lands about seven or eight minutes later and clears out of the system.

"At 9:30 the Langley F-16s are airborne. They are 105 miles away from the Washington area. (It was later determined, however, rather than heading north toward Washington, the fighters were flying east over the Atlantic. When they finally got turned around, they were sent toward Baltimore, thinking they would protect the Capitol from attack from the north.)[9]

"9:34, through chat, FAA is unable to precisely locate American Airlines Flight 77.

"9:35, F-16s are reported airborne. And many times, reported airborne is not exactly when they took off. It's just when the report came down that they were airborne.

"At 9:37 we have the last radar data near the Pentagon. [The F-16s from Langley had just turned around and were still 150 miles away from Washington.]

"And 9:40, immediately following that, is when 93 up north turns its transponders off out in the west toward Ohio, and begins a left turn back toward the East. . . .

"9:47 is when Delta 89 clears the system by landing in Cleveland. So he is not a hijack. . . .

"At 10:02, United 93 last radar data and the estimated impact time for United 93 is 10:03.

"At 10:07 FAA reports there may be a bomb on board 93—that's four minutes after the impact.

"At 10:15 they report that it's crashed. And you can see now that fog and friction lag time has increased from seven minutes to nine minutes to fifteen minutes, because of the level of activities that are going on. . . . We're picking up the phone, calling Syracuse, the Air National Guard. They're beginning to get flights airborne. They're beginning to arm those aircraft with whatever weapons they have handy so we can posture that defense. That is how the timeline unfolded."

President George Bush that day was in Florida doing campaign-style events at local schools. Orders were given in his stead by Vice President Dick Cheney midway through the hijacking spree to defend the nation's capital at all costs, including shooting down hijacked airliners. Transportation Secretary Norman Mineta, whose agency oversaw civil aviation, became aware of this almost accidentally. He had rushed to the White House and been escorted to a secure bunker from which Cheney was operating. He arrived just as United 93 was thought to be bearing down on Washington. The Secret Service was projecting its flight path from its last known position. They had no actual radar data. An aide was calling out its progress to the vice president:

"'The plane is 50 miles out. The plane is 30 miles out.' And when it got down to, 'The plane is 10 miles out,' the young man also said to the vice president, 'Do the orders still stand?'

"And the vice president turned, and whipped his neck around and said, 'Of course the orders still stand. Have you heard anything to the contrary?'"[10]

The aircraft had long since crashed in Pennsylvania. As NORAD noted in its reconstruction, fighter jets were scrambled all over the

Northeast and mid-Atlantic coast. The only aircraft in a position in time to defend actual targets that day were two F-16 fighters from Langley. Although orders had been given on the ground allowing the fighters to shoot down any hijacked airliners they encountered, the orders never made it to the F-16s.

As blundering as the air defense system looked in the NORAD reconstruction, in actuality, it was even worse. The scrambled fighters from Langley that NORAD had said were sent aloft when they believed American 77 was headed for Washington were actually sent in the mistaken belief that American 11, the first aircraft to hit the World Trade Center, had somehow kept flying south past New York toward Washington. The Langley F-16s were sent to attack a plane that had crashed long before they received their first order.[11]

The four targeted flights had been scheduled to take off within fifteen minutes of one another to prevent exactly the sort of problem that occurred on Flight 93—feedback from the ground leaking back onto the flights. Flight 93 was in the air nearly two hours after Flight 77 hit the Trade Center. Passengers and crew, alerted by people on the ground to what had already happened that day, staged a revolt on board that eventually led the hit team to crash the flight headfirst into the Pennsylvannia countryside rather than yield it to their captives.[12] Flight 93 passengers armed with cell phones and pitchers of hot water were able to effectively defend the nation's capital in a way the national air defense system could not.

That day, in the air, inside the four aircraft, we know now there was near pandemonium. There was war. Omar expressed the correct attitude of the attackers by quoting a Muslim scholar, Abdul Abbas al-Zahrani: "I have come to you with an imminent slaughter."

And so they did. The knives were used. It was violent and bloody and thick. The slaughter was more than imminent; it was realized, fully.

AFTER

By volume, most of the two towers of the World Trade Center were empty air. All big buildings are, but the Trade Center was especially so. The late John Skilling, the head of the structural engineering firm that designed the towers, used to enjoy showing a chart of all the lightest tall buildings in the world. In a yellow band at the top of the chart were the lightest buildings ever built. His designs were clustered in that band; as a group, they were overwhelmingly the lightest tall buildings ever erected. He achieved this extraordinary lightness mainly by reducing the amount of steel in the buildings, creating more space for tenants and much more money for landlords, since space equates to rent and since structural steel was typically the single most expensive element of any big building. Many of these buildings are 1 million square feet or more (each of the Trade Center towers was a gargantuan 4.3 million square feet), so even the slightest reduction of steel per square foot was amplified in builders' bank accounts. In the era when the World Trade Center was designed, the weights for high-rises often exceeded 75 pounds of steel per square foot. The World Trade Center was 37 pounds, saving tens of millions of dollars.

Making tall buildings weigh less has effects beyond the balance sheet. There are fundamentally two types of forces to be addressed in making sure a building stands up: a vertical force, gravity, which seeks to pull it down; and lateral forces, generally wind or earthquakes, which seek to push it over. The effects of gravity, because they are constant, are easier to predict and overcome and are somewhat lessened by reducing the weight of the buildings. The lateral forces, however, which vary unpredictably, present a greater challenge to an engineer. Because they were so light, the main structural concern with Skilling's buildings was always wind. At a glance, it seems unlikely that anything weighing many millions of pounds would be susceptible to the wind, but the overall density of many of Skilling's buildings was equivalent to that of balsa wood. Consider what the wind might do to a 700-foot-tall piece of balsa.

The Trade Center towers were, in fact, almost willowy. They would move 10 feet at the top in a heavy wind. They could have been engineered safely to move even more. Exactly how much sway to allow was a novel problem the builders had to address and its solution was based on creature comfort, not engineering skill. Skilling's firm and architect Minoru Yamasaki set out to determine at what level of acceleration—it is the speed of the movement, not the amount, that human beings react to—an occupant felt a building move. To measure this, they built a room atop a hydraulic pad and advertised free eye exams. People came in for their exams and sat in a chair, and while they deciphered the vision charts, psychologists, hidden behind two-way mirrors, ordered the room to be moved "in a figure eight, which is how a building moves."[13]

Reactions varied dramatically. Some people complained at the slightest acceleration. Others gripped their chairs as if they were riding a Tilt-A-Whirl, yet never uttered a murmur of protest. Skilling and the psychologists eventually established "a threshold of awareness. We determined the rate at which 2 percent of the people seriously objected and set that as the upper limit."

The Trade Center towers had to have some ability to sway. If they didn't bend, they would stress and eventually break, and they were designed not to break even in the face of 140-mile-per-hour gusts of hurricane-force wind. But the inertial force of wind isn't shaped into the sharp point of a very large arrow moving at 500 miles per hour; even hurricanes don't attack at the speed of a Boeing 767. Eerily, another of the buildings' engineers once bragged that the towers were designed to withstand the impact of an airliner; people laughed when they heard that anyone would ever consider such a thing.

In the event, the buildings performed as expected. They handily absorbed the impacts. In effect, they caught the airplanes, quivered, and leaned and—holes torn in their sides, bleeding smoke and hot metal—returned in just a few seconds to their upright stance. By that day, Skilling had long since passed away. One of his former partners, Jon Magnusson, was watching on television when the second plane, Marwan al-Shehhi's, drove into the south tower. Magnusson was, of course, appalled, but proud, too.[14] The twin towers took the mightiest of blows and stood, still. They could not absorb what followed.

After Atta hit the north tower and Shehhi followed into the south, 10,000 gallons of kerosene were ignited inside each building with the destructive force of 7 million sticks of dynamite, lighting ferocious fires of more than 1,300 degrees that eventually melted steel, buckling columns and collapsing floors, one on top of another, until first the south tower, then the north, crumbled and pancaked to the ground, rendering into a seven-story-tall stack of rubble the giant buildings and their contents, a million tons of glass, stone, steel, paper, concrete, plastic, polished mahogany desks, Crane's 24-bond embossed letterhead stationery, janitors' mops, caterers' donuts and coffee carts, Brioni's finest three-button all-weather wool suits, Kiton English twill-silk seven-fold ties, Herman Miller Aeron chairs, complicated causes of legal actions, high-speed computers—far-flung hopes, dreams, and memories all gone in the premature demise of 2,792 suddenly, but completely, vanquished lives.

Appendix A

Mohamed el-Amir's Last Will and Testament

In the name of God all mighty
Death Certificate
This is what I want to happen after my death, I am Mohamed the son of Mohamed Elamir awad Elsayed: I believe that prophet Mohamed is God's messenger and time will come no doubt about that and God will resurrect people who are in their graves. I wanted my family and everyone who reads this will to fear the Almighty God and don't get deceived by what is in life and to fear God and to follow God and his prophets if they are real believers. In my memory, I want them to do what Ibrahim [a prophet] told his son to do, to die as a good Muslim. When I die, I want the people who will inherit my possessions to do the following:

1. The people who will prepare my body should be good Muslims because this will remind me of God and his forgiveness.
2. The people who are preparing my body should close my eyes and pray that I will go to heaven and to get me new clothes, not the ones I died in.
3. I don't want anyone to weep and cry or to rip their clothes or slap their faces because this is an ignorant thing to do.
4. I don't want anyone to visit me who didn't get along with me while I was alive or to kiss me or say good bye when I die.

5. I don't want a pregnant woman or a person who is not clean to come and say good bye to me because I don't approve it.

6. I don't want women to come to my house to apologize for my death. I am not responsible for people who will sacrifice animals in front of my lying body because this is against Islam.

7. Those who will sit beside my body must remember Allah, God, and pray for me to be with the angels.

8. The people who will clean my body should be good Muslims and I do not want a lot of people to wash my body unless it is necessary.

9. The person who will wash my body near my genitals must wear gloves on his hands so he won't touch my genitals.

10. I want the clothes I wear to consist of three white pieces of cloth, not to be made of silk or expensive material.

11. I don't want any women to go to my grave at all during my funeral or on any occasion thereafter.

12. During my funeral I want everyone to be quiet because God mentioned that he likes being quiet on occasions when you recite the Qur'an, during the funeral, and when you are crawling. You must speed my funeral procession and I would like many people there to pray for me.

13. When you bury me the people with whom I will be buried should be good Muslims. I want to face East toward Mecca.

14. I should be laying on my right side. You should throw the dust on my body three times while saying from the dust, we created you dust and to dust you will return. From the dust a new person will be created. After that everyone should mention God's name and that I died as a Muslim which is God's religion. Everyone who attends my funeral should ask that I will be forgiven for what I have done in the past (not this action).

15. The people who will attend my funeral should sit at my grave for an hour so that I will enjoy their company and slaughter animals and give the meat to the needy.

16. The custom has been to memorialize the dead every forty days or once a year but I do not want this because it is not an Islamic custom.

17. I don't want people to take time to write things on paper to be kept in their pockets as superstition. Time should be taken to pray to God instead.

18. All the money I left must be divided according to the Muslim religion as almighty God has asked us to do. A third of my money should be spent on the poor and the needy. I want my books to go to any one of the Muslim mosques. I wanted the people who look at my will to be one of the heads of the Sunna religion. Whoever it is, I want that person to be from where I grew up or any person I used to follow in prayer. People will be held responsible for not following the Muslim religion. I wanted the people who I left behind to hear God and not to be deceived by what life has to offer and to pray more to God and to be good believers. Whoever neglects this will or does not follow the religion, that person will be held responsible in the end.

This was written on April 11, 1996, the Islamic calendar of *ʒoelqada* is 1416.

Written by Mohamed Mohamed Elamir Awad Elsayed

Witness: Abdelghani Mouzdi

Witness: Mounir el-Motassadeq

Appendix B

The Last Night

A PORTION OF THE INSTRUCTIONS for the hijack teams. It is believed to have been written by Abdul Aziz al-Omari.

1. Vow to accept death, renew admonition, shave the extra hair on the body, perfume yourself, and ritually wash yourself.
2. Know the plan well from every angle. Anticipate the reaction or the resistance of the enemy.
3. Read the surahs of Repentance and The Spoils. Contemplate their meaning and the bounties God has prepared and established for the martyrs.
4. Remind your base self to listen and obey this night, for you will be exposed to decisive turning points wherein listening and obeying is one hundred percent necessary. Train your base self, make it understand, convince it, and goad it on to this end. "And obey God, and His Messenger, and do not quarrel together, and so lose heart, and your power depart; and be patient; surely God is with the patient."
5. Staying up at night and imploring in prayer for victory and strength and perspicuous triumph, and the easing of our task, and concealment.
6. Much recitation of sacred phrases. Know that the best of *dhikr* is reciting the noble Qur'an. This is the consensus of

the people of knowledge or, indeed, of the most learned. It is enough for us that it is the words of the creator of the heavens and the earth toward Whom you are advancing.

7. Purify your heart and cleanse it of stains. Forget and be oblivious to that thing called the world. For, the time for playing has passed, and the time has arrived for the rendezvous with the eternal Truth. How much of our lives we have wasted! Shall we not take advantage of these hours to offer up acts of nearness [to God] and obedience?

8. Let your breast be filled with gladness, for there is nothing between you and your wedding but mere seconds. Thereby will begin a happy and contented life and immortal blessing with the prophets, the true ones and the righteous martyrs. They are the best of companions. We beseech God for his grace. So seek good omens. For the Prophet, may blessings and peace be upon him, used to love divination about every matter.

9. Then fix your gaze, such that if you fall into tribulations, you will know how to behave, how to stand firm, how to say "We are, verily, from God and to him we shall return." Thus you will know that what has befallen you is not because of any error you committed. That you committed an error was not so that you would face tribulations. That calamity of yours is in fact from God, may he be exalted and glorified—so as to elevate your station and cause your sins to be forgiven. Know that it is only a matter of seconds before it shines forth by the permission of God. Then blessed is he who attains the great recompense from God. God says, "Did you think you would enter paradise when God knows those who strove among you, and knows the patient?"

10. Then recite the words of God, "You were wishing for death before you encountered it, then you saw it, and are looking for it." And you wanted it. After then, recite the verse

"Kam min fi'ah qalilah ghalaba fi'ah kathirah bi idhn Allah." And "In yunsirukum Allah fa la ghalib lakum."

11. Bring your base self, as well as your brethren, to remembrance through prayers. And contemplate their meaning (recitations [*adhkar*] of morning and evening, recitations of city [*baldah*], recitations of . . . [*makan*], recitations of meeting [*liqa' al-Tur*]. . . .

12. The jet: suitcase, clothing, knife, tools, identity papers, passport, and all your papers.

13. Inspect your weapon before setting out and before you even begin to set out. . . .

14. Pull your clothes tightly about you, for this is the way of the pious ancestors (*as-salaf as-salih*), may God be pleased with them. They pulled their clothing tightly about them before a battle. Pull your shoelaces tight and wear tight socks that grip the shoes and do not come out of them. All of these are means that we have been commanded to adopt. God has *hasabna* and he is the best of advocates (*na'im al-wakil*).

15. Pray the morning prayers in congregation and reflect on the reward for doing so while you are performing recitation afterwards. Do not go out of your apartment without having performed ablutions. For the angels seek forgiveness for you as long as you have prepared ablutions and they pray on your behalf.

Appendix C

Bin Laden's 1996 Declaration of War Against the Americans
Occupying the Land of the Two Holy Places (abridged)

P RAISE BE TO ALLAH, we seek His help and ask for his pardon. We
take refuge in Allah from our wrongs and bad deeds. Who ever been guided
by Allah will not be misled, and who ever has been misled, he will never
be guided. I bear witness that there is no God except Allah—no associates with Him—and I bear witness that Muhammad is His slave and
messenger. . . .

It should not be hidden from you that the people of Islam had suf-
fered from aggression, iniquity and injustice imposed on them by the
Zionist-Crusaders alliance and their collaborators; to the extent that the
Muslims blood became the cheapest and their wealth as loot in the hands of
the enemies. Their blood was spilled in Palestine and Iraq. The horrifying
pictures of the massacre of Qana, in Lebanon are still fresh in our memory.
Massacres in Tajikistan, Burma, Kashmir, Assam, Philippine, Fatani,
Ogadin, Somalia, Eritrea, Chechnya and in Bosnia-Herzegovina took
place, massacres that send shivers in the body and shake the conscience.
All of this and the world watch and hear, and not only didn't respond to
these atrocities, but also with a clear conspiracy between the USA and
its' allies and under the cover of the iniquitous United Nations, the dis-
possessed people were even prevented from obtaining arms to defend
themselves.

The people of Islam awakened and realized that they are the main
target for the aggression of the Zionist-Crusaders alliance. All false claims

and propaganda about "Human Rights" were hammered down and exposed by the massacres that took place against the Muslims in every part of the world.

The latest and the greatest of these aggressions, incurred by the Muslims since the death of the Prophet (ALLAH'S BLESSING AND SALUTATIONS ON HIM) is the occupation of the land of the two Holy Places—the foundation of the house of Islam, the place of the revelation, the source of the message and the place of the noble Ka'ba, the Qiblah of all Muslims—by the armies of the American Crusaders and their allies. (We bemoan this and can only say: "No power and power acquiring except through Allah.")

Under the present circumstances, and under the banner of the blessed awakening which is sweeping the world in general and the Islamic world in particular, I meet with you today. And after a long absence, imposed on the scholars (*Ulema*) and callers (*Da'ees*) of Islam by the iniquitous crusaders movement under the leadership of the USA; who fears that they, the scholars and callers of Islam, will instigate the Ummah of Islam against its' enemies as their ancestor scholars—may Allah be pleased with them—like Ibn Taymiyyah and Al'iz Ibn Abdes-Salaam did. And therefore the Zionist-Crusader alliance resorted to killing and arresting the truthful Ulema and the working Da'ees. (We are not praising or sanctifying them; Allah sanctify whom He pleased.) They killed the Mujahid Sheikh Abdullah Azzam, and they arrested the Mujahid Sheikh Ahmad Yaseen and the Mujahid Sheikh Omar Abdur Rahman (in America).

By orders from the USA they also arrested a large number of scholars, Da'ees and young people—in the land of the two Holy Places—among them the prominent Sheikh Salman Al-Oud'a and Sheikh Safar Al-Hawali and their brothers. (We bemoan this and can only say: "No power and power acquiring except through Allah.") We, myself and my group, have suffered some of this injustice ourselves; we have been prevented from addressing the Muslims. We have been pursued in Pakistan, Sudan and Afghanistan, hence this long absence on my part. But by the Grace of Allah, a safe base is now available in the high Hindu Kush mountains in Khurasan; where—by the Grace of Allah—the largest infidel military force of the world was destroyed. And the myth of the super power was withered in front of the Mujahideen cries of Allahu Akbar (God is greater). Today we work from the same mountains to lift the iniquity that had been imposed on the Ummah by the Zionist-Crusader alliance, particularly after they have occupied the blessed land around Jerusalem, route of the journey of the Prophet (ALLAH'S BLESSING AND SALUTATIONS ON HIM) and the land of the two Holy Places. We ask Allah to bestow us with victory, He is our Patron and He is the Most Capable.

From here, today we begin the work, talking and discussing the ways of correcting what had happened to the Islamic world in general, and the Land of the two Holy Places in particular. We wish to study the means that we could follow to return the situation to its' normal path. And to return to the people their own rights, particularly after the large damages and the great aggression on the life and the religion of the people. An injustice that had affected every section and group of the people; the civilians, military and security men, government officials and merchants, the young and the old people as well as schools and university students. Hundred of thousands of the unemployed graduates, who became the widest section of the society, were also affected. . . .

The explosion at Riyadh and Al-Khobar is a warning of this volcanic eruption emerging as a result of the severe oppression, suffering, excessive iniquity, humiliation and poverty. . . .

Through its course of actions the regime has torn off its legitimacy:

(1) Suspension of the Islamic Sharia law and exchanging it with man made civil law. The regime entered into a bloody confrontation with the truthful Ulema and the righteous youths (we sanctify nobody; Allah sanctify Whom He pleaseth).

(2) The inability of the regime to protect the country, and allowing the enemy of the Ummah—the American crusader forces—to occupy the land for the longest of years. The crusader forces became the main cause of our disastrous condition, particularly in the economical aspect of it due to the unjustified heavy spending on these forces. As a result of the policy imposed on the country, especially in the field of oil industry where production is restricted or expanded and prices are fixed to suit the American

economy ignoring the economy of the country. Expensive deals were imposed on the country to purchase arms. People asking what is the justification for the very existence of the regime then? . . .

The right answer is to follow what have been decided by the people of knowledge, as was said by Ibn Taymiyyah (Allah's mercy upon him): "people of Islam should join forces and support each other to get rid of the main 'Kufr' who is controlling the countries of the Islamic world, even to bear the lesser damage to get rid of the major one, that is the great Kufr."

If there are more than one duty to be carried out, then the most important one should receive priority. Clearly after Belief (Imaan) there is no more important duty than pushing the American enemy out of the holy land. No other priority, except Belief, could be considered before it; the people of knowledge, Ibn Taymiyyah, stated: "to fight in defense of religion and Belief is a collective duty; there is no other duty after Belief than fighting the enemy who is corrupting the life and the religion. There is no preconditions for this duty and the enemy should be fought with one best abilities (ref: supplement of Fatawa). If it is not possible to push back the enemy except by the collective movement of the Muslim people, then there is a duty on the Muslims to ignore the minor differences among themselves; the ill effect of ignoring these differences, at a given period of time, is much less than the ill effect of the occupation of the Muslims' land by the main Kufr. Ibn Taymiyyah had explained this issue and emphasized the importance of dealing with the major threat on the expense of the minor one. He described the situation of the Muslims and the Mujahideen and stated that even the military personnel who are not practicing Islam are not exempted from the duty of Jihad against the enemy. . . .

Utmost effort should be made to prepare and instigate the Ummah against the enemy, the American-Israeli alliance—occupying the country of the two Holy Places and the route of the Apostle (Allah's Blessings and Salutations may be on him) to the Furthest Mosque (Al-Aqsa Mosque). Also to remind the Muslims not to be engaged in an internal war among themselves, as that will have grieve consequences namely:

1. Consumption of the Muslims human resources as most casualties and fatalities will be among the Muslims people.

2. Exhaustion of the economic and financial resources.

3. Destruction of the country infrastructures

4. Dissociation of the society

5. Destruction of the oil industries. The presence of the USA Crusader military forces on land, sea and air of the states of the Islamic Gulf is the greatest danger threatening the largest oil reserve in the world. The existence of these forces in the area will provoke the people of the country and induces aggression on their religion, feelings and prides and push them to take up armed struggle against the invaders occupying the land; therefore spread of the fighting in the region will expose the oil wealth to the danger of being burned up. The economic interests of the States of the Gulf and the land of the two Holy Places will be damaged and even a greater damage will be caused to the economy of the world. I would like here to alert my brothers, the Mujahideen, the sons of the nation, to protect this (oil) wealth and not to include it in the battle as it is a great Islamic wealth and a large economical power essential for the soon to be established Islamic state, by Allah's Permission and Grace. We also warn the aggressors, the USA, against burning this Islamic wealth (a crime which they may commit in order to prevent it, at the end of the war, from falling in the hands of its legitimate owners and to cause economic damages to the competitors of the USA in Europe or the Far East, particularly Japan which is the major consumer of the oil of the region).

6. Division of the land of the two Holy Places, and annexing of the northerly part of it by Israel. Dividing the land of the two Holy Places is an essential demand of the Zionist-Crusader alliance. The existence of such a large country with its huge resources under the leadership of the forthcoming Islamic State, by Allah's Grace, represent a serious danger to the very existence of the Zionist state in Palestine. The Nobel Ka'ba,—the Qiblah of all Muslims—makes the land of the two Holy Places a symbol for the unity of the Islamic

world. Moreover, the presence of the world largest oil reserve makes the land of the two Holy Places an important economical power in the Islamic world. The sons of the two Holy Places are directly related to the life style (Seerah) of their forefathers, the companions, may Allah be pleased with them. They consider the Seerah of their forefathers as a source and an example for re-establishing the greatness of this Ummah and to raise the word of Allah again. Furthermore the presence of a population of fighters in the south of Yemen, fighting in the cause of Allah, is a strategic threat to the Zionist-Crusader alliance in the area. The Prophet (ALLAH'S BLESSING AND SALUTATIONS ON HIM) said: (around twelve thousands will emerge from Aden/Abian helping—the cause of—Allah and His messenger, they are the best, in the time, between me and them) narrated by Ahmad with a correct trustworthy reference.

7. An internal war is a great mistake, no matter what reasons are there for it. The presence of the occupier—the USA— forces will control the outcome of the battle for the benefit of the international Kufr. . . .

It is out of date and no longer acceptable to claim that the presence of the crusaders is necessity and only a temporary measures to protect the land of the two Holy Places. Especially when the civil and the military infrastructures of Iraq were savagely destroyed showing the depth of the Zionist-Crusaders hatred to the Muslims and their children, and the rejection of the idea of replacing the crusaders forces by an Islamic force composed of the sons of the country and other Muslim people. Moreover the foundations of the claim and the claim it self were demolished and wiped out by the sequence of speeches given by the leaders of the Kufr in America. The latest of these speeches was the one given by William Perry, the Defense Secretary, after the explosion in Al-Khobar saying that: the presence of the American soldiers there is to protect the interest of the USA. The imprisoned Sheikh Safar Al-Hawali, may Allah hasten his release, wrote a book of seventy pages; in it he presented evidence and proof that the presence of the Americans in the Arab Peninsula is a pre-planed military occupation. The regime want to deceive the Muslim people in the same manner when the Palestinian fighters, Mujahideen, were deceived causing the loss of Al-Aqsa Mosque. In 1304 A.H. (1936 AD) the awakened Muslims nation of Palestine started their great struggle, Jihad, against the British occupying forces. Britain was impotent to stop the Mujahideen and their Jihad, but their devil inspired that there is no way to stop the armed struggle in Palestine unless through their agent King Abdul Azeez, who managed to deceive the Mujahideen. King Abdul Azeez carried out his duty to his British masters. He sent his two sons to meet the Mujahideen leaders and to inform them that King Abdul Azeez would guarantee the promises made by the British government in leaving the area and responding positively to the demands of the Mujahideen if the latter stop their Jihad. And so King Abdul Azeez caused the loss of the first Qiblah of the Muslims people. The King joined the crusaders against the Muslims and instead of supporting the Mujahideen in the cause of Allah, to liberate the Al-Aqsa Mosque, he disappointed and humiliated them.

Today, his son, King Fahd, trying to deceive the Muslims for the second time so as to lose what is left of the sanctities. When the Islamic world resented the arrival of the crusader forces to the land of the two Holy Places, the king told lies to the Ulema (who issued Fatwas about the arrival of the Americans) and to the gathering of the Islamic leaders at the conference of Rabitah which was held in the Holy City of Mecca. The King said that: "the issue is simple, the American and the alliance forces will leave the area in few months." Today it is seven years since their arrival and the regime is not able to move them out of the country. The regime made no confession about its inability and carried on lying to the people claiming that the American will leave. But never-never again; a believer will not be bitten twice from the same hole or snake! Happy is the one who takes note of the sad experience of the others!!

Instead of motivating the army, the guards, and the security men to oppose the occupiers, the regime used these men to protect the invaders, and further deepening the humiliation and the betrayal. (We bemoan this and can only say: "No power and power acquiring except through Allah.") To those little group of men within the army, police and security forces, who have been tricked and pressured by the regime to attack the Muslims and spill their blood, we would like to remind them of the narration: (I promise war against those who take my friends as their enemy) narrated by Al-Bukhari. And his saying (Allah's Blessings and Salutations may be on him) saying of: (In the day of judgment a man comes holding another and complaining being slain by him. Allah, blessed be His Names, asks: Why did you slay him?! The accused replies: I did so that all exaltation may be Yours. Allah, blessed be His Names, says: All exaltation is indeed mine! Another man comes holding a fourth with a similar complaint. Allah, blessed be His Names, asks: Why did you kill him?! The accused replies: I did so that exaltation may be for Mr. X! Allah, blessed be His Names, says: exaltation is mine, not for Mr. X, carry all the slain man's sins (and proceed to the Hell fire!). In another wording of An-Nasa'i: "The accused says: for strengthening the rule or kingdom of Mr. X."

Today your brothers and sons, the sons of the two Holy Places, have started their Jihad in the cause of Allah, to expel the occupying enemy from of the country of the two Holy places. And there is no doubt you would like to carry out this mission too, in order to re-establish the greatness of this Ummah and to liberate its' occupied sanctities. Nevertheless, it must be obvious to you that, due to the imbalance of power between our armed forces and the enemy forces, a suitable means of fighting must be adopted, i.e. using fast moving light forces that work under complete secrecy. In other word to initiate a guerrilla warfare, where the sons of the nation, and not the military forces, take part in it. And as you know, it is wise, in the present circumstances, for the armed military forces not to be engaged in a conventional fighting with the forces of the crusader enemy (the exceptions are the bold and the forceful operations carried out by the members of the armed forces individually, that is without the movement of the formal forces in its conventional shape and hence the responses will not be directed, strongly, against the army) unless a big advantage is likely to be achieved; and great losses induced on the enemy side (that would shake and destroy its foundations and infrastructures) that will help to expel the defeated enemy from the country.

The Mujahideen, your brothers and sons, requesting that you support them in every possible way by supplying them with the necessary information, materials and arms. Security men are especially asked to cover up for the Mujahideen and to assist them as much as possible against the occupying enemy; and to spread rumors, fear and discouragement among the members of the enemy forces.

We bring to your attention that the regime, in order to create a friction and feud between the Mujahideen and yourselves, might resort to take a deliberate action against personnel of the security, guards and military forces and blame the Mujahideen for these actions. The regime should not be allowed to have such opportunity.

The regime is fully responsible for what had been incurred by the country and the nation; however the occupying American enemy is the principle and the main cause of the situation . Therefore efforts should be concentrated on destroying, fighting and killing the enemy until, by the Grace of Allah, it is completely defeated. The time will come—by the Permission of Allah—when you'll perform your decisive role so that the word of Allah will be supreme and the word of the infidels (Kaferoon) will be the inferior. You will hit with iron fist against the aggressors. You'll re-establish the normal course and give the people their rights and carry out your truly Islamic duty. Allah willing, I'll have a separate talk about these issues.

My Muslim Brothers (particularly those of the Arab Peninsula): The money you pay to buy American goods will be transformed into bullets and used against our brothers in Palestine and tomorrow (future) against our sons in the land of the two Holy places. By buying these goods we are strengthening their economy while our dispossession and poverty increases.

Muslims Brothers of land of the two Holy Places:

It is incredible that our country is the world largest buyer of arms from the USA and the area biggest commercial partners of the Americans who are assisting their Zionist brothers in occupying Palestine and in evicting and killing the Muslims there, by providing arms, men and financial supports.

To deny these occupiers from the enormous revenues of their trading with our country is a very important help for our Jihad against them. To express our anger and hate to them is a very important moral gesture. By doing so we would have taken part in (the process of) cleansing our sanctities from the crusaders and the Zionists and forcing them, by the Permission of Allah, to leave disappointed and defeated.

We expect the woman of the land of the two Holy Places and other countries to carry out their role in boycotting the American goods.

If economical boycotting is intertwined with the military operations of the Mujahideen, then defeating the enemy will be even nearer, by the Permission of Allah. However if Muslims don't co-operate and support their Mujahideen brothers then, in effect, they are supplying the army of the enemy with financial help and extending the war and increasing the suffering of the Muslims.

The security and the intelligence services of the entire world can not force a single citizen to buy the goods of his/her enemy. Economical boycotting of the American goods is a very effective weapon of hitting and weakening the enemy, and it is not under the control of the security forces of the regime. . . .

Few days ago the news agencies had reported that the Defense Secretary of the Crusading Americans had said that "the explosion at Riyadh and Al-Khobar had taught him one lesson: that is not to withdraw when attacked by coward terrorists."

We say to the Defense Secretary that his talk can induce a grieving mother to laughter! and shows the fears that had enshrined you all. Where was this false courage of yours when the explosion in Beirut took place on 1983 AD (1403 A.H.). You were turned into scattered pits and pieces at that time; 241 mainly marines soldiers were killed. And where was this courage of yours when two explosions made you to leave Aden in less than twenty four hours!

But your most disgraceful case was in Somalia; where—after vigorous propaganda about the power of the USA and its post cold war leadership of the new world order—you moved tens of thousands of international force, including twenty eight thousands American soldiers into Somalia. However, when tens of your soldiers were killed in minor battles and one American Pilot was dragged in the streets of Mogadishu you left the area carrying disappointment, humiliation, defeat and your dead with you. Clinton appeared in front of the whole world threatening and promising revenge, but these threats were merely a preparation for withdrawal. You have been disgraced by Allah and you withdrew; the extent of your impotence and weaknesses became very clear. It was a pleasure for the "heart" of every Muslim and a remedy to the "chests" of believing nations to see you defeated in the three Islamic cities of Beirut, Aden and Mogadishu.

I say to Secretary of Defense: The sons of the land of the two Holy Places had come out to fight against the Russian in Afghanistan, the Serb in Bosnia-Herzegovina and today they are fighting in Chechnya and—by the Permission of Allah—they have been made victorious over your partner, the Russians. By the command of Allah, they are also fighting in Tajikistan.

I say: Since the sons of the land of the two Holy Places feel and strongly believe that fighting (Jihad) against the Kufr in every part of the world, is absolutely essential; then they would be even more enthusiastic, more powerful and larger in number upon fighting on their own land— the place of their births—defending the greatest of their sanctities, the noble Ka'ba (the Qiblah of all Muslims). They know that the Muslims of the world will assist and help them to victory. To liberate their sanctities is the greatest of issues concerning all Muslims; it is the duty of every Muslims in this world.

I say to you William (Defense Secretary) that: These youths love death as you loves life. They inherit dignity, pride, courage, generosity, truthfulness and sacrifice from father to father. They are most delivering and steadfast at war. They inherit these values from their ancestors (even from the time of the Jaheliyyah, before Islam). These values were approved and completed by the arriving Islam as stated by the messenger of Allah (Allah's Blessings and Salutations may be on him): "I have been send to perfecting the good values" (Saheeh Al-Jame' As-Sagheer).

When the pagan King Amroo Ibn Hind tried to humiliate the pagan Amroo Ibn Kulthoom, the latter cut the head of the King with his sword, rejecting aggression, humiliation and indignation. . . .

Those youths know that their rewards in fighting you, the USA, is double than their rewards in fighting some one else not from the people of the book. They have no intention except to enter paradise by killing you. An infidel, and enemy of God like you, cannot be in the same hell with his righteous executioner. . . .

So to abuse the grandsons of the companions, may Allah be pleased with them, by calling them cowards and challenging them by refusing to leave the land of the two Holy Places shows the insanity and the imbalance you are suffering from. Its appropriate "remedy," however, is in the hands of the youths of Islam, as the poet said:

I am willing to sacrifice self and wealth for knights who never disappointed me.

Knights who are never fed up or deterred by death, even if the mill of war turns.

In the heat of battle they do not care, and cure the insanity of the enemy by their "insane" courage.

Terrorizing you, while you are carrying arms on our land, is a legitimate and morally demanded duty. It is a legitimate right well known to all humans and other creatures. Your example and our example is like a snake which entered into a house of a man and got killed by him. The coward is the one who lets you walk, while carrying arms, freely on his land and provides you with peace and security.

Those youths are different from your soldiers. Your problem will be how to convince your troops to fight, while our problem will be how to restrain our youths to wait for their turn in fighting and in operations. . . .

I remind the youths of the Islamic world, who fought in Afghanistan and Bosnia-Herzegovina with their wealth, pens, tongues and themselves that the battle had not finished yet. I remind them about the talk between Jibreel (Gabriel) and the messenger of Allah (Allah's Blessings and Salutations may be on both of them) after the battle of Ahzab when the messenger of Allah (Allah's Blessings and Salutations may be on him) returned to Medina and before putting his sword aside; when Jibreel (Allah's Blessings and Salutations may be on him) descend saying: "are you putting your sword aside? by Allah the angels haven't dropped their arms yet; march with your companions to Bani Quraydah, I am (going) ahead of you to throw fears in their hearts and to shake their fortresses on them." Jibreel marched with the angels (Allah's Blessings and Salutations may be on them all), followed by the messenger of Allah (Allah's Blessings and Salutations may be on him) marching with the immigrants, Muhajeroon, and supporters, Ansar (narrated by Al-Bukhary).

These youths know that: if one is not to be killed one will die (any way) and the most honorable death is to be killed in the way of Allah. They are even more determined after the martyrdom of the four heroes who bombed the Americans in Riyadh. Those youths who raised high the head of the Ummah and humiliated the Americans—the occupier—by their operation in Riyadh. They remember the poetry of Ja'far, the second commander in the battle of Mu'tah, in which three thousand Muslims faced over a hundred thousand Romans:

How good is the Paradise and its nearness, good with cool drink. But the Romans are promised punishment (in Hell), if I meet them.

I will fight them. . . .

My Muslim Brothers of The World:

Your brothers in Palestine and in the land of the two Holy Places are calling upon your help and asking you to take part in fighting against the enemy—your enemy and their enemy—the Americans and the Israelis. They are asking you to do whatever you can, with one own means and ability, to expel the enemy, humiliated and defeated, out of the sanctities of Islam. . . .

Appendix D

Bin Laden's 1998 Jihad Against Jews and Crusaders

World Islamic Front Statement
23 February 1998
Signators:
Shaykh Usamah Bin-Muhammad Bin-Ladin
Ayman al-Zawahiri, amir of the Jihad Group in Egypt
Abu-Yasir Rifa'i Ahmad Taha, Egyptian Islamic Group
Shaykh Mir Hamzah, secretary of the Jamiat-ul-Ulema-e-
    Pakistan
Fazlur Rahman, amir of the Jihad Movement in Bangladesh

PRAISE BE TO ALLAH, who revealed the Book, controls the clouds, defeats factionalism, and says in His Book: "But when the forbidden months are past, then fight and slay the pagans wherever ye find them, seize them, beleaguer them, and lie in wait for them in every stratagem (of war)"; and peace be upon our Prophet, Muhammad Bin-'Abdallah, who said: I have been sent with the sword between my hands to ensure that no one but Allah is worshipped, Allah who put my livelihood under the shadow of my spear and who inflicts humiliation and scorn on those who disobey my orders.

The Arabian Peninsula has never—since Allah made it flat, created its desert, and encircled it with seas—been stormed by any forces like the crusader armies spreading in it like locusts, eating its riches and wiping out its plantations. All this is happening at a time in which nations are attacking Muslims like people fighting over a plate of food. In the light of the grave situation and the lack of support, we and you are obliged to discuss current events, and we should all agree on how to settle the matter.

No one argues today about three facts that are known to everyone; we will list them, in order to remind everyone:

First, for over seven years the United States has been occupying the lands of Islam in the holiest of places, the Arabian Peninsula, plundering its riches, dictating to its rulers, humiliating its people, terrorizing its neighbors, and turning its bases in the Peninsula into a spearhead through which to fight the neighboring Muslim peoples.

If some people have in the past argued about the fact of the occupation, all the people of the Peninsula have now acknowledged it. The best proof of this is the Americans' continuing aggression against the Iraqi people using the Peninsula as a staging post, even though all its rulers are again their territories being used to that end, but they are helpless.

Second, despite the great devastation inflicted on the Iraqi people by the crusader-Zionist alliance, and despite the huge number of those killed, which has exceeded 1 million . . . despite all this, the Americans are once again trying to repeat the horrific massacres, as though they are not content with the protracted blockade imposed after the ferocious war or the fragmentation and devastation.

So here they come to annihilate what is left of this people and to humiliate their Muslim neighbors.

Third, if the Americans' aims behind these wars are religious and economic, the aim is also to serve the Jews' petty state and divert attention from its occupation of Jerusalem and murder of Muslims there. The best proof of this is their eagerness to destroy Iraq, the strongest neighboring Arab state, and their endeavor to fragment all the states of the region such as Iraq, Saudi Arabia, Egypt, and Sudan into paper statelets and through their disunion and weakness to guarantee Israel's survival and the continuation of the brutal crusade occupation of the Peninsula.

All these crimes and sins committed by the Americans are a clear declaration of war on Allah, his messenger, and Muslims. And ulema have throughout Islamic history unanimously agreed that the jihad is an individual duty if the enemy destroys the Muslim countries. This was revealed by Imam Bin-Qadamah in "Al-Mughni," Imam al-Kisa'i in "Al-Bada'i," al-Qurtubi in his interpretation, and the sheikh of al-Islam in his books, where he said: "As for the fighting to repulse [an enemy], it is aimed at defending sanctity and religion, and it is a duty as agreed [by the ulema]. Nothing is more sacred than belief except repulsing an enemy who is attacking religion and life."

On that basis, and in compliance with Allah's order, we issue the following fatwa to all Muslims:

The ruling to kill the Americans and their allies—civilians and military—is an individual duty for every Muslim who can do it in any country in which it is possible to do it, in order to liberate the al-Aqsa Mosque and the holy mosque [Mecca] from their grip, and in order for their armies to move out of all the lands of Islam, defeated and unable to threaten any Muslim. This is in accordance with the words of Almighty Allah, "and fight the pagans all together as they fight you all together," and "fight them until there is no more tumult or oppression, and there prevail justice and faith in Allah."

This is in addition to the words of Almighty Allah: "And why should ye not fight in the cause of Allah and of those who, being weak, are ill-treated (and oppressed)?—women and children, whose cry is: 'Our Lord, rescue us from this town, whose people are oppressors; and raise for us from thee one who will help!'"

We—with Allah's help—call on every Muslim who believes in Allah and wishes to be rewarded to comply with Allah's order to kill the Americans and plunder their money wherever and whenever they find it. We also call on Muslim ulema, leaders, youths, and soldiers to launch the raid on Satan's U.S. troops and the devil's supporters allying with them, and to displace those who are behind them so that they may learn a lesson.

Almighty Allah said: "O ye who believe, give your response to Allah and His Apostle, when He calleth you to that which will give you life. And know that Allah cometh between a man and his heart, and that it is He to whom ye shall all be gathered."

Almighty Allah also says: "O ye who believe, what is the matter with you, that when ye are asked to go forth in the cause of Allah, ye cling so heavily to the earth! Do ye prefer the life of this world to the hereafter? But little is the comfort of this life, as compared with the hereafter. Unless ye go forth, He will punish you with a grievous penalty, and put others in your place; but Him ye would not harm in the least. For Allah hath power over all things."

Almighty Allah also says: "So lose no heart, nor fall into despair. For ye must gain mastery if ye are true in faith."

Notes

U NLESS OTHERWISE NOTED or attributed, all quotations used in the book are derived from interviews conducted between 2001 and 2004 by the author or colleagues cited in the Acknowledgments. Quotations used in reconstructed conversations at which the author was obviously not present are derived from author interviews or police interrogations of at least one of the direct participants in the quoted conversations and are noted throughout. Additional direct quotations from principal figures in the attacks are derived from investigative records in which those principals are quoted and are noted throughout.

PROLOGUE: WELCOME
1. As quoted in a sermon at Al·Quds that was videotaped and later recovered by German investigators. Date unknown, circa 2000.
2. Ibid.
3. German law-enforcement source, author interview, Hamburg, February 2005.
4. Ibid.

Book One: Soldiers

CHAPTER 1: THE HOUSE OF LEARNING
1. Gerhard Rotter, interview by author, Hamburg, November 2001.
2. Hamida Fateh, interview by author, Kafr el-Sheik, October 2001.
3. Ayman al-Zawahiri, "Knights Under the Prophet's Banner," *Al Sharq Al Awsat*, December 2001. In this memoir, serialized in the London-based newspaper, Zawahiri, Osama Bin Laden's deputy and the leader of the Egyptian group Al-Jihad, gives an insider's account of the underground politics of Cairo in these years.
4. There is an extensive academic and historical literature on the origins of fundamentalism and its political role in the Middle East. An accessible summary is provided by David Zeidan, "The Islamic Fundamentalist View of Life as a Perennial

Battle," *Middle East Review of International Affairs,* Volume 5, no. 4, December 2001. The Princeton University historian Bernard Lewis has written extensively on the subject. He presents his view in, *The Crisis of Islam: Holy War and Unholy Terror,* Modern Library, March 2003. Also excellent is *The Shade of Swords,* by M. J. Akbar, Routledge, 2002.

5. Cairo's professional ranks have formed cooperative societies called syndicates, through which members obtain insurance, special rates for vacations, education programs, and so forth. It has been reported repeatedly that Amir joined the Engineer's Syndicate, which in addition to the social benefits conferred on members, was at the time controlled by Cairo's Muslim Brotherhood. This makes sense, but a thorough search of the syndicate's membership records failed to turn up any membership, either before or after Amir was graduated from university.

CHAPTER 2: ALONE, ABROAD

1. Host couple, interview by author, Hamburg, December 2001.
2. Ibid.
3. Rüdiger Bendlin, TUHH administrator, interview by author, Harburg, November 2001.
4. Dittmar Machule, interview by author, Harburg, November 2001.
5. Hans Harms, interview by author, Harburg, November 2001.
6. Martin Ebert, telephone interview by author, December 2001.
7. Harmut Kaiser, telephone interview by author, December 2001.
8. Amir's first and second roommates, interviews by author, Germany, November 2001.
9. Manfred Schröder, interview by author, Harburg, November 2001.
10. From the investigative files of the Bundeskriminalamt, or BKA, the German equivalent of the FBI, author copy, July 2002.
11. Volker Hauth, interview by author, Hamburg, November 2001.
12. Roommate's girlfriend, interview by author, Hamburg, November 2001.
13. Machule, interview, op. cit.
14. Razan Abdel-Wahab, interview by Michael Slackman for the *Los Angeles Times,* November 2001.
15. Hauth, interview, op. cit.
16. Ibid.
17. Mohamed el-Amir, interview by author, Cairo, October 2001.
18. Hauth, interview, op. cit.
19. Ralph Bodenstein, telephone interview by author, October 2001.
20. Nader Fergany, "Human Capital And Economic Performance In Egypt," Almishkat Centre for Research, Cairo, August 1998.
21. Hauth, interview, op. cit.

22. John Sadiq, correspondence to the author, December 2001.
23. Jörg Lewin, interview by author, Hamburg, November 2001.
24. Helga Rake, interview by author, Hamburg, November 2001.

CHAPTER 3: FRIENDS

1. TUHH records, author copy.
2. Hauth, interview, op. cit.
3. Shahid Nickels, BKA interrogations, November 2001, author copy.
4. Office for the Protectors of the Constitution, *Annual Report, 2000,* Berlin.
5. Ibid.
6. Nickels, interrogations, op cit.
7. Ahmed Maklat, BKA interrogations, November 2001, author copy.
8. Shahid Nickels, interview by author, Hamburg, September 2004.
9. Ibid.
10. Maklat, interrogations, op. cit.
11. Ibid.
12. Nickels, interview, op. cit.
13. Ramzi bin al-Shibh, Auslander, or German Foreigners Bureau investigative report, author copy.
14. Ibid.
15. Egyptian police records, quoted by Auslander investigators, author copy.
16. Mohamed Ahmed bin al-Shibh, Omar's brother, interview by author, Sana'a, August 2003.
17. Faris Sanabani, adviser to the president of Yemen, Ali Abdullah Saleh, interview by author, Sana'a, August 2003.
18. Ahmed bin al-Shibh, op. cit.
19. Bakir el-Beihi, mosque student, interview by author, Sana'a, August 2003.
20. Bernard Haykel, a scholar who specializes in Islamism at New York University, said Islam is so deeply embedded in Yemeni culture as to be nearly genetic. Even moderate, secular Yemenis, he said, "might not pray but they would be willing to die for Islam."
21. Most of these men were native Saudis, including some who would become notorious; for example, Abd al Rahim al Nashiri, who would become a key Al Qaeda leader, was one of those welcomed to settle in Yemen.
22. Hassam Sanabani, interview by author, Sana'a, August 2003.
23. BKA federal investigative report, author copy, October 2001.
24. Michael Hirsekorn, camp director, interview by author, Pinneberg, Germany, March 2002.
25. German federal investigative files, author copy.

26. Volker Harun Bruhn, Belfas's friend, interview by author, April 2002.
27. Nickels, interrogations, op. cit.
28. Maklat, interrogations, op. cit.
29. German immigration files, author copy.
30. Marion Koll, spokeswoman for the Court of Schleswig, author interview by telephone, Hamburg, April 2002.
31. Ibid.
32. Nickels, interview, op. cit. There was always "some business going on," Nickels said.
33. German welfare records, author copy, October 2001. German investigators have been able to reconstruct some of Shibh's spending habits using the food coupons given the asylum seekers.
34. Amir's first roommate, interview, op. cit.

CHAPTER 4: PILGRIMS

1. Lewin, interview, op. cit.
2. Seminar attendee, interview by author, Hamburg, December 2001.
3. Nickels, interrogation, op. cit.
4. Nickels, interview, op. cit.
5. Ibid.
6. Much has been made of the fact that Amir, then a mere twenty-eight years old, had executed a last will and testament, but it was far from uncommon among young Muslim men. Templates for them were passed around the mosques and are still. Motassadeq also wrote a will about the same time. Others downloaded templates for jihad wills from the Internet and followed suit. One man instructed his survivors to pay all his debts to Muslims but withhold all money from Christians and Jews. This isn't to say that some of the content of Amir's will isn't odd. A translation of the will is included in Appendix A.
7. Senior German investigator, interview by author, Hamburg, July 2002. The bookstore was under surveillance by German investigators and many of the conversations were surreptitiously recorded. The recordings were made later, after Amir and Omar's group had dissolved, but they provide good insight into the overwhelming role religion played for young men in these situations.
8. Senior German investigator, interview by author, Hamburg, April 2002.
9. Palestinians were a political as well as a physical presence in the life of Jarrah's neighborhood. In an interview with the author, Assem Jarrah, a cousin, said he attended university on a Palestinian Liberation Organization scholarship.
10. Ziad Jarrah high school transcripts, author copy, August 2001.
11. Jamal Jarrah, interview by author, Al Marj, Lebanon, August 2003.

12. Aysel Sengün, BKA interrogation, author copy, November 2001. Unless otherwise noted, all the material in this account of the personal relationship between Sengün and Jarrah is drawn from the BKA files or Aysel's testimony in the trial of Mounir el-Motassadeq.

13. Assem Jarrah, Ziad's cousin, who preceded him to Griefswald by several years and lived there while Ziad was there, interview by author, Beirut, August 2003.

14. Salim Jarrah, interview by author, Griefswald, April 2002.

15. Siri Kannengiesser, Aysel Sengün's Greifswald roommate, quoted in BKA investigative files, November 2001, author copy.

16. Abdulrachman al-Makhadi, who disclaims anything but the slightest acquaintance with Ziad Jarrah, seems the likeliest candidate to have influenced him. Investigators at some point had begun monitoring Makhadi, whom they classified as "an endangerer" of other Muslims. In an interview with the author, Makhadi lied vigorously about his relationship with Jarrah, which investigative records indicate grew quite close. They made weekend trips together from Greifswald, including at least one to the western town of Aachen, which for more than a decade had been a center of radical Islam in Europe. But beyond their association investigators say they have no proof of what, if any, influence Makhadi had over Jarrah.

17. Aysel Sengün and Ziad Jarrah, BKA files of letters and e-mail between the two, author copy, various dates 1999–2001.

18. Sengün, interrogation, op. cit.

19. Nickels, interrogations, op. cit.

20. Maklat, interrogations, op. cit.

21. UAE government officials provided the basic outline of his biography, which is confirmed and detailed in BKA files.

22. Mohamed Abdullah Mohamed Awady and Hamad Marzooqi, quoted in BKA investigative files, November 2001, author copy.

23. Marzooqi, Ibid.

24. Awady, op. cit.

25. There is little in the investigative record that sheds light on Shehhi's desire to move. Most simply, he could easily have gleaned enough information from the Islamist grapevine to conclude Hamburg offered a more hospitable climate for him. It also would not have been at all far-fetched for him to have met some of the Hamburg men. Omar, for example, traveled often to Bonn and attended services at the same mosque Shehhi attended.

26. Amir's second roommate, interview, op. cit.

27. BKA files, author copy. Auslander files, October 2001. Also, CIA director

George Tenet buttressed this view in his testimony before Congress in June 2003.

28. Shahid Nickels, testimony at second trial of Mounir el-Motassadeq, Hamburg, August 17, 2004.

29. The chronologies of the men were developed primarily from credit, bank, and university records included in the BKA files, supplemented with police interrogation reports and interview by author of acquaintances.

30. Owner of Hay Computing, interview by author, Hamburg, December 2001.

31. Manfred Schröder, interview by author, Harburg, November 2001.

32. Helga Link, interview by author, Hamburg, November 2001.

33. Nickels, testimony, second Motassadeq trial, op. cit.

34. Maklat, interrogations, op. cit.

35. Nickels, interview, op. cit.

36. Berlin woman, testimony at second trial of Mounir el-Motassadeq, Hamburg, August 18, 2004. The woman was allowed to testify in secret without revealing her identity. She said Omar wanted to marry, but she felt he was too religious. She told the court of one ritual she found odd: for his daily morning prayer, he always wore the same pair of white trousers and his Yemeni dagger, which he unfailingly gripped during the prayer.

37. Nickels, interview, op. cit.

38. Maklat, interrogations, op. cit.

39. Ibid.

40. Nickels, interview, op. cit.

41. Volker Harun Bruhn, interview, op. cit.

42. Nickels, interview, op. cit.

43. Nickels, interrogations, op. cit.

44. Nickels, testimony, second Motassadeq trial, op. cit.

45. Said Bahaji, BKA files, October 2001.

46. BKA files, October 2001.

47. Thorsten Albrecht, interview by author, Hamburg, November 2001.

48. Owner of firm that employed Bahaji, interview by author, Hamburg, June 2002.

49. Bahaji loved Michael Schumacher, the dominant Formula One driver and the biggest sports star in Germany by far. Bahaji had Schumacher's posters on his wall and watched the races through a video card he installed on his computer. This was not entirely unusual among the members of the Hamburg group. Another man plastered the door of his student apartment with photos of Ferraris from car-enthusiast magazines.

50. Barbara Arens, interview by author, Hamburg, May 2002.

51. Nickels, interrogations, op. cit.

52. Bahaji remained angry at her, however. He later wrote extremely vitriolic letters to her. The BKA recovered crates of materials from Bahaji's apartment and thousands of files from his computer, including his correspondence, from which these details are drawn.
53. Nickels, interrogation, op. cit.
54. Yassir Boughlal, BKA interrogation, November 2001, author copy.
55. Maklat, interrogations, op. cit.
56. Yassir Boughlal, testimony at first trial of Mounir el-Motassadeq, Hamburg, October 2002.
57. Ibid.
58. Ibid. Ironically, even though they were often targets of Islamist discontent, Western liberal democracies were much more hospitable to Islamists than were their home countries.
59. Nickels, interrogations, op. cit.
60. Ahmed Maklat, testimony at first trial of Mounir el-Motassadeq, October 2002.
61. Ibid.
62. Ibid.
63. Nickels, testimony, second Motassadeq trial, op. cit.
64. Nickels, interview, op. cit.
65. Maklat, interrogations, op. cit.
66. Nickels, testimony, second Motassadeq trial, op. cit.

CHAPTER 5: THE SMELL OF PARADISE RISING

1. Unless otherwise noted, this account is drawn from interviews with German and American investigators, BKA investigative files, and court records and trial transcripts of various American Al Qaeda prosecutions, especially *USA* v *Usama bin Laden et al.*, United States District Court, Southern District of New York, 2001.
2. House Permanent Select Committee on Intelligence and the Senate Select Committee on Intelligence, *Report of the Joint Inquiry Committee into the Terrorist Attacks of September 11, 2001,* December 2002; hereafter cited as the Congressional Joint Inquiry Committee.
3. Mamdouh Salim is still in American custody. He was charged, along with Bin Laden and others, in the embassy bombing case, but his case was delayed after he attacked and nearly killed a guard with a sharpened comb while in federal detention in New York City. He subsequently pleaded guilty to attempted murder of the guard and has yet to come to trial in the bombing case.
4. Senior German investigator, interview by author, Hamburg, November 2001.
5. Nickels, interrogations, op. cit.

6. BKA files, author copy. Bahaji's computer was later inspected by German investigators. The hard drive was filled with Islamist materials.
7. Abdel Dahuk, interview by author, Hamburg, April 2002.
8. Mohammed Mohammed Zammar, interview by author, Hamburg, April 2002. Mohammed said that his brother Haydar had joined the Tabligh several years prior after an invitation to meet with the group in Pakistan.
9. Oliver Schröm and Dirk Laabs, "The Deadly Mistakes of the U.S. Intelligence Agency," *Stern*, August 13, 2003.
10. Congressional Joint Inquiry Committee, *Report*.
11. BKA files, author copy. The outline of this information was reconfirmed by CIA director George Tenet in his testimony before both the Congressional Joint Inquiry Committee and the separate 9/11 Commission. Tenet dismissed the usefulness of the information, saying all his agency was given was a first name and an impossible-to-trace unlisted telephone number in the UAE. At least a portion of that statement is preposterous. The UAE mobile telephone business was, until 2004, a state monopoly. The UAE number could have been traced in five minutes, according to senior security officials there. The United States never asked. Of equal curiosity is why the Germans would have given the information to the Americans in the first place. The content of the intercepted phone call was innocuous. Shehhi called Zammar. After an exchange of pleasantries, this is what was said, all in paraphrase by German investigators.

Zammar: I heard you were back in Bonn.
Shehhi: Yes, and how was your trip to Syria?
Zammar: Good.
Shehhi: I'm living with a friend now in Bonn, but will move into a student apartment next week. My studies are continuing.
Zammar: I'm unemployed, but I may get new job training next year.
Zammar: Marwan, you should try to come in March.
Shehhi: Do you want to travel?
Zammar: No.
Shehhi: I will have tests beginning in February. Once all tests are done in May, I will come to Hamburg and take a flat.
Zammar: Are you married yet?
Shehhi: Not exactly.
Zammar: How's your brother, Mohammed?
Shehhi: He's fine. Greetings to Abula Fata [a cousin of Zammar's] and Abu Ilyas [Darkazanli].

This conversation does not seem to be of any particular import, so why pass it on? One plausible answer is that the Americans had a standing request for any new information about Zammar. Neither the Germans nor the Americans would comment on that. But a previous interest in Zammar would explain, among other things, why Thomas Volz, the CIA agent, was so adamant about trying to turn Darkazanli into an informer. Further, the CIA told the Joint Inquiry Committee it had a longstanding interest in Zammar that pre-dated the period of these recordings. In other words, the CIA appears to have been investigating the man who recruited the hijackers at the time he was recruiting them.

12. Senior German intelligence officer, interview by author, Hamburg, November 2001.

13. This activity was not restricted to Europe. A Pakistani expatriate in the United Arab Emirates told how he and a group of fellow religionists were recruited by members of the Tabligh for a journey of spiritual renewal to Pakistan, and when they arrived found themselves in the middle of a firefight between Pakistanis and Indians in Kashmir. When the men objected to the danger there, they were transported clandestinely to a Tabligh camp in western Pakistan, where Al Qaeda representatives made repeated efforts to recruit them; interview by author, UAE, July 2003.

14. Mohammed Mohammed Zammar, interview, op. cit.

15. German investigative records, author copies.

16. Nickels, testimony at first trial of Mounir el-Motassadeq, Hamburg, October 2002.

17. Congressional Joint Inquiry Committee, p. 185. Darkazanli was arrested in late 2004.

18. Schröm and Laabs, "Deadly Mistakes."

19. Unbeknown to Volz or his CIA superiors, even if he might have been susceptible to persuasion, money was not much of a problem for Darkazanli. Later investigations determined that hundreds of thousands of dollars flowed through his accounts, much of it from Saudi Arabia.

20. Machule, interview, op. cit.

21. Ibid.

22. Quoted in BKA files, August 31, 1999, author copy.

23. In general, remaining single is frowned upon within Islam. Yusuf al-Qaradawi, a Qatari cleric popular among the Hamburg group, has written, in "The Halal and the Haram in Marriage," that Muslims are compelled to marry: "As long as he possesses the means to marry, the Muslim is not permitted to refrain from marriage on the grounds that he has dedicated himself to the service or the worship of Allah and to a life of monasticism and renunciation of the world."

24. Sengün, interrogations, op. cit.

25. Jarrah, e-mails contained in BKA investigative files, author copy, various dates 1999–2001.

26. Sengün, interrogations, op. cit.

27. Ibid.

28. Sengün, e-mails, quoted in BKA records, author copy, various dates 1999–2001.

29. Ibid.

30. Siri Kannengiesser, quoted in BKA investigative files, op. cit.

31. Ramzi bin al-Shibh, Hamburg, October 1999. The wedding was videotaped with a home video camera. These quotes and descriptions are from the author's copy of that video.

32. Mohamed el-Amir, interview by author, Cairo, October 2001.

33. Fateh, interview, op. cit.

34. BKA investigative files, author copy.

35. Ahmed Hashim, "The World According to Osama Bin Laden," *Naval War College Review*, Autumn 2001, p. 18.

36. Nickels, interrogations, op. cit.

37. Osama Bin Laden, "Declaration of War Against the Americans Occupying the Land of the Two Holy Places," *Al Quds Al Arabi*, August 1996. An abridged version appears in Appendix C.

38. Nickels, interview, op. cit.

39. Nickels, interrogations, op. cit. The obvious contradiction between this live-and-let-live philosophy and the notion of waging a holy war against your religious enemies is neatly elided in the Islamist literature by making a distinction between the individual, whose opinion you must respect, and states, whose only purpose is to keep believers from living a true Islamic life. You may, and in fact must, according to Islamist doctrine, attack states and their representatives even while respecting the rights of individuals to disagree with you.

40. Volker Harun Bruhn, interview by author, Hamburg, May 2002.

41. German law-enforcement source, author interview, Hamburg, February 2005.

42. Ibid.

43. These dated notes, a kind of rough journal, were found among Jarrah's university papers and are contained in BKA investigative files, author copy, October 2001.

44. BKA investigative files, author copy.

45. Nickels, interrogations, op. cit.

Book Two: The Engineer
CHAPTER 1: THE REBIRTH OF JIHAD

1. Abdullah Azzam, "Defense of Muslim Lands." Accessed at http://www.religioscope.com/info/doc/jihad/azzam-defence_6_chap1.htm.
2. Abdullah Azzam, *Join the Caravan*, Azzam Publications, London, 2001, p. 9.
3. Azzam, "Defense of Muslim Lands," op. cit.
4. Barnett R. Rubin provides an excellent history of modern Afghanistan in *The Fragmentation of Afghanistan*, 2d ed., Yale University Press, New Haven, 2002.
5. Ibid, p. 179.
6. Ibid, p. 199.
7. Brzezinski later boasted in an interview about having tricked the Soviets into Afghanistan where he envisioned, correctly as it happened, they would be bogged down for years: "That secret operation was an excellent idea. It had the effect of drawing the Russians into the Afghan trap and you want me to regret it? The day that the Soviets officially crossed the border, I wrote to President Carter: We now have the opportunity of giving to the USSR its Vietnam war. Indeed, for almost 10 years, Moscow had to carry on a war unsupportable by the government, a conflict that brought about the demoralization and finally the breakup of the Soviet empire." See *Le Nouvel Observateur* (France), January 15–21, 1998, p. 76.
8. Waheed Hamza Hashim, interview by author, Jeddah, Saudi Arabia, August 2003.
9. Mohammad Yousaf, a former ISI officer writing with Mark Adkin, chronicles the Pakistani efforts in *Afghanistan the Bear Trap: The Defeat of a Superpower*, Casemate Publishers, Havertown, Pa., 1992. Accessed at http://www.afghanbooks.com/beartrap/english/18.htm.
10. Ahmed Rashid, *Taliban*, Pan Books, London, 2001, p. 129.
11. The shrine contracts lasted decades and solidified the company's influence with the royal family. The Saudi Bin Laden Group continues to be among the largest privately owned enterprises in the entire Middle East.
12. Jane Mayer, "The House of Bin Laden," *The New Yorker*, November 12, 2001.
13. Bernard Haykel, telephone interview by author, February 2004.
14. Bernard Haykel, "Radical Salafism," *Dawn*, Pakistan, December 8, 2001.
15. Zawahiri, "Knights Under the Prophet's Banner," op. cit.
16. Ibid.
17. A few camps, run by the Pakistani military, had nothing to do with Afghanistan at all, but existed solely to train fighters for Pakistan's other war—with India in Kashmir.
18. Friends of Sayyaf, interviews by author, Kabul, November 2003.
19. Amin Farhang, interview by author, Kabul, November 2003.

20. Robert Eastham, telephone interview by author, January 2004.

21. Ibid.

22. Milton Bearden, telephone interview by author, December 2003.

CHAPTER 2: THOSE WITHOUT

1. Senior Kuwaiti security service officer, interview by author, Kuwait City, August 2003.

2. "Baluchistan During Persian Empires." Accessed at http://www.Baluch 2000.org.

3. Hashim Abdul Aziz, interview by author, Fahaheel, August 2003.

4. Friend of Mohammed brothers, interview by Josh Meyer for the *Los Angeles Times*, Islamabad, Pakistan, November 2002.

5. Shafeeq Ghabra, interview by Patrick McDonnell for the *Los Angeles Times*, Kuwait City, November 2002.

6. Simon Reeve, *The New Jackals*, Northeastern University Press, 1999, p. 113. Reeve's book focuses on Ramzi Yousef and Osama Bin Laden and is the best-researched and most accurate of the many books on the West's initial encounters with Islamic terrorism. Its material on Yousef, in particular, remains without parallel.

7. Ibid.

8. The National Commission on Terrorist Attacks Upon the United States, *Final Report*, July 22, 2004, p. 145; also known as the 9/11 Commission and hereafter cited as the National Commission.

9. National Commission, *Staff Statement 16*, June 16, 2004.

10. Badawi Hindieh, telephone interview by Patrick McDonnell for the *Los Angeles Times*, November 2002. McDonnell's reporting on Khalid's college years was unsurpassed and forms the basis for much of this section.

11. Descriptions of Khalid Sheikh Mohammed's life in North Carolina are based primarily on research and interviews conducted there by Patrick McDonnell for the *Los Angeles Times*, October 2002.

12. Ibid.

13. Garth D. Faile, chairman of the Chowan science department, interview by Patrick McDonnell, October 2002.

14. Khalil A. Abdullah, telephone interview by Patrick McDonnell, October 2002.

15. Quentin Clay, telephone interview by author, September 2002.

16. Sami Zitawi, telephone interview by author, September 2002.

17. Mahmood Zubaid, interview by Patrick McDonnell, Kuwait City, November 2002.

18. Waleed M. Qimlass, interview by Patrick McDonnell, Kuwait City, November 2002.

19. Zubaid, interview, op. cit.

20. UAE official, interview by author, United Arab Emirates, August 2003.

21. Mohammed Zubaid, interview by author, Kuwait City, August 2003.

22. Faisal Munifi, interview by Patrick McDonnell, Kuwait City, November 2002.

23. Zitawi, interview, op. cit.

24. Hindieh, interview by Patrick McDonnell, op. cit.

25. Kuwaiti official, interview by author, op. cit.

26. Mohammed's teacher, interview by author, op. cit.

27. Annual reports filed and records maintained by the Islamic Coordinating Committee, Peshawar.

28. Reeve, *The New Jackals*, op. cit., p. 48.

29. Friend of Mohammed brothers, interview by Josh Meyer, op. cit.

30. Ibid.

31. Former employee of Muslim relief agencies in Peshawar, interviewed by Josh Meyer for the *Los Angeles Times*, Islamabad, November 2002.

32. Shir Alam, interview by author, Kabul, November 2003.

33. Ibid.

34. Burhanuddin Rabbani, interview by author, Kabul, November 2003.

35. Ahmed Wali Massoud, interview by author, Kabul, November 2003. Massoud was among the most respected of the Afghan military commanders and one of those most resistant to the sort of fundamentalist Islam embraced by those commanders favored by the United States and Saudi Arabia. Ironically, he later became the Americans' strongest ally in confronting the Taliban and Osama Bin Laden. He was assassinated by Bin Laden operatives on September 9, 2001.

36. Robert Eastham, telephone interview by author, January 2004.

37. Reeve, *The New Jackals*, p. 164.

38. International Crisis Group, "Jemaah Islamiyah in South East Asia: Damaged But Still Dangerous," Asia Report No. 3, August 26, 2003, p. 4. This remarkably thorough ICG report is based on interviews with JI prisoners and written primarily by ICG analyst Sidney Jones, who has since been deported from Indonesia largely because of the thoroughness and accuracy of her research.

39. Peter Bergen, *Holy War, Inc.*, Touchstone, New York, 2002, p. 60.

40. Reeve, *The New Jackals*, p. 165.

41. Steve Coll, *Ghost Wars: The Secret History of the CIA, Afghanistan, and Bin Laden, from the Soviet Invasion to September 10, 2001*, Penguin Press, New York, 2004, p. 163.

42. John K. Cooley, *Unholy Wars: Afghanistan, America, and International Terrorism*, Pluto Press, London, 2000, p. 109. See also, Coll, *Ghost Wars*, p. 149. Coll provides the most comprehensive, detailed history of the American involvement.

43. Coll, *Ghost Wars*, p. 158.

44. Even as the Soviets left, the United States refused to believe they were going. The Soviets had repeatedly contacted the Americans, trying to initiate peace negotiations, but the Americans thought it was a trick. Coll offers an authoritative account.

45. Robert Oakley, telephone interview by author, January 2004.

46. Milt Bearden, telephone interview by author, January 2004.

47. Rubin, *Fragmentation*, p. 250.

48. Coll, *Ghost Wars*, pp. 195–197.

49. Ojhri was an ammunition dump that was blown up outside Islamabad not long before the Jalalabad battle. It was the ISI's main source of supply. Ten thousand tons of ordnance were lost, including 30,000 rockets, thousands of mortar bombs, and millions of rounds of small-arms ammunition. See Yousaf and Adkin, *Afghanistan the Bear Trap*.

50. Abdullah Azzam, "Lover of the Paradise Maidens," as quoted in the *Muslim News*, London, May 3, 2003.

51. Coll, *Ghost Wars*, pp. 195–199.

52. Rubin, *Fragmentation*, p. 251.

53. Jamal Ahmed al-Fadl, testimony in *USA v Usama bin Laden et al.*, February 6, 2001. Fadl's testimony is the basic source for almost everything that has been written about the early days of Al Qaeda. He had been one of the first members of the organization, but later defected and became a key U.S. government informant against Bin Laden.

54. There is considerable debate as to when Bin Laden began to use the term Al Qaeda to refer to his organization. The word is used in Arabic to mean a base, a foundation, or a pillar. There are many who think that it was just in this colloquial sense that the term was used early on, to refer to function not to an actual organization. Nonetheless, a variation of it—*Qa'idat al-Jihad*, or The Base of the Jihad—over time came to be the term commonly used to refer to Bin Laden and his hard-core supporters, who never numbered more than several hundred. Jason Burke, *Al Qaeda: Casting a Shadow of Terror*, Tauris, London, 2003, pp. 7–22, devotes an entire chapter to a lucid discussion of the term and the organization. Burke's book is excellent generally; its more moderate tone is in great contrast to much else that has been written. The book benefits greatly from Burke's decade of reporting in the region.

55. The Prophet Mohammed had commanded: "Expel the polytheists from the Arabian Peninsula." This has been taken by Bin Laden to mean that no one but pure Muslims can live in Saudi Arabia.

56. Coll, *Ghost Wars*, pp. 221–223.

CHAPTER 3: WORLD WAR

1. Richard Clarke, chief of counter-terrorism in the Clinton and Bush administrations, interviewed March 2002, Accessed at http://www.pbs.org/wgbh/pages/frontline/shows/knew/interviews/clarke.html.

2. Simon Reeve, *The New Jackals*, p. 115. Reeve relies on Pakistani investigators for this information. Kuwaiti authorities confirmed the Pakistani account in interviews, and Abdul Basit himself listed a home address in Baluchistan when he married. It might be, however, that the family moved to Iranian Baluchistan rather than Pakistani Baluchistan. Yousef told an acquaintance interviewed by the author that his wife and children lived in Iran. The two regions, though divided by a national border, were virtually indistinguishable, and people moved back and forth between the two as if they were one. Many Baluch nationalists did not recognize the supremacy of any government other than their own. Abdul Basit's father, Abdul Karim, was thought to count himself among those nationalists.

3. Reeve, *The New Jackals*. Reeve does not name the brothers, but later events indicate the two cited were allied with the resistance.

4. Steve Coll, *Ghost Wars*, p. 248.

5. Wedding certificate.

6. Brian Parr, testimony in *USA* v *Yousef, et al.*, S1293CR.180, August 12, 1996, author transcript.

7. Former jihadis, interviews by author, Peshawar, November 2002, and Dubai, UAE, August 2003.

8. Edwin Angeles, former Abu Sayyaf member, quoted in "Philippine Jihad Inc.," unpublished report by Sr. Supt. Rodolfo Mendoza Jr., Philippine National Police, September 2002, author copy.

9. International Crisis Group, "Jemaah Islamiyah in South East Asia," p. 2.

10. Abdul Hakim Murad, interrogation by FBI, April 1995, author copy.

11. Ibid.

12. Jamal Ahmed al-Fadl, testimony in *USA* v *Usama bin Laden et al.*, S(7)98Cr.1023, February 20, 2001.

13. Coll, *Ghost Wars*, op. cit, pp. 221–224.

14. Friend of Mohammed brothers, interview by Josh Meyer, op. cit.

15. National Commission, *Final Report*, July 22, 2004, p. 147.

16. Abdul Hakim Murad, interrogation by Philippine National Police, January, 1995, author copy.

17. Ibid.

18. Ahmad Mohammed Ajaj's biographical information is contained in the trial transcripts of *USA* v *Salameh, et al*, S593CR.180

19. This, at least, was the story Ajaj's lawyer told at trial, that he was simply trying

to get back to the United States and was offered the ticket by Abdul Basit; and other than the fact that he was traveling with Basit, there was very little evidence offered to suggest he was part of any plan to attack the United States. He was in jail at the time of the bomb attack (in fact, during the entire time Yousef was in the United States.), but was charged in the conspiracy nonetheless.

20. All of the details about Yousef's entry into the United States are drawn from transcripts of *USA* v *Salameh et al.*, *USA* v *Yousef et al.*, and *USA* v *Ismail et al.*, author copies. Many of the details are repeated in the different trials and are scattered throughout.

21. Abdul Basit, interrogation by FBI, February 1995, author copy.

22. Ibid.

23. *USA* v *Yousef*, op. cit.

24. This rather remarkable line of work was first reported by Simon Reeve in *The New Jackals*. At the time, Reeve did not know Magid's identity, but investigators later confirmed that he was, in fact, Khalid Sheikh Mohammed. They also confirmed in interviews the holy water story, but it still seems somewhat unlikely. I do not intend to cast doubt on Reeve's work. His is by far the best of the pre-9/11 books on Islamist terror. But there is an echo chamber effect in cases like this that persists over many years. In that time, investigators forget where they learned things and are not always diligent about sorting fact from rumor. They might well have read the holy water story in Reeve's book, which is where I first saw it, or in the Pakistani press, where it was reported—then repeated it without ever ascertaining its credibility. I've often heard investigators repeat to me things I know they knew from having read them in a newspaper story I wrote. I doubt they fact-checked my reporting.

25. National Commission, *Final Report*, July 22, 2004, p. 148.

26. Ibid, p. 147.

27. All the information about Murad's relationship with Yousef comes from transcripts of Murad's interrogations by Philippine intelligence, January 1995, and summaries of his interrogation by the FBI, April 1995, author copies.

28. ABC News interview with Osama Bin Laden, May 28, 1998. Accessed at http://abcnews.go.com/sections/world/DailyNews/miller_binladen_980609.html.

29. Reeve, *The New Jackals*, pp. 52–54.

30. Ibid, pp. 63–65.

31. Mior Mohamad Yuhana, interview by author, February 2002.

32. International Crisis Group, "Indonesia Backgrounder: How the Jemaah Islamiyah Terrorist Network Operates," Asia Report No. 43, December 11, 2002.

33. "The Islamic Fundamentalist/Extremist Movements in the Philippines and

Their Links With International Terrorist Organizations," unpublished special report by the Philippines National Police Intelligence Command, Special Investigations Group, December 1994, author copy.

34. Ibid.

35. Singapore Government White Paper on Jemaah Islamiyah, January 2002.

36. Yousef was an incorrigible flirt. Later, when he was on trial in New York, he tried to get a date with one of the court stenographers.

37. Maria Ressa, *Seeds of Terror,* Free Press, New York, 2004, pp. 21–24.

38. Sr. Supt. Rodolfo Mendoza Jr., interview by author, August 2002.

39. Angeles, op. cit.

40. Murad, interrogations, op. cit.

41. The neighborhood has been cleaned up since, but residents complain that the politicians who did it didn't know what they were doing. The reformed neighborhood is becalmed but boring.

42. Michael Garcia, deputy U.S. attorney, quoted in *USA* v *Yousef et al.,* trial transcripts, author copy.

43. Murad, interrogations, op. cit.

44. "The Islamic Fundamentalist/Extremist Movements in the Philippines and Their Links with International Terrorist Organizations," an unpublished special report by the Philippines Intelligence Command Special Investigations Group, December 1994, author copy.

45. Senior Philippine intelligence officers, interviews by author, Manila, August 2002.

46. Mendoza, "Philippine Jihad Inc.," op. cit.

47. Jose L. Atienza Jr., mayor of Manila, interview by author, Manila, August 2002.

48. Murad, interrogations, op. cit.

49. Ibid. The plans, complete with flight numbers and code names for the bombers, were found on Yousef's computer, along with a trove of other incriminating documents.

50. Government investigators, already suspicious of the men in room 603, and worried about the pope's imminent arrival, surreptitiously set off the alarm, according to Jose L. Atienza Jr., who was then the city councilman for the district of Malate; interview, op. cit.

51. Philippine National Police inventory of evidence seized, author copy.

52. Ibid.

53. Mendoza, interview, op. cit.

54. Robin Wright and John-Thor Dahlburg, "Legwork, Luck Closed Net Around Bombing Suspect," *Los Angeles Times,* February 12, 1995.

CHAPTER 4: WAR, AFTER WAR

1. Jamal Ahmed al-Fadl, *USA* v. *Bin Laden*, February 6, 2001. S(7)98CR.1023.
2. Ibid.
3. Steve Coll, *Ghost Wars*, p. 222.
4. "Usama Bin Laden: Islamic Extremist Financier," unpublished report of the CIA, 1996, author copy.
5. Ibid.
6. Cofer Black, former CIA counterterrorism director, testifying before the Congressional Joint Inquiry Committee, September 2003.
7. American diplomat, interview by author, Washington, D.C., October 2002.
8. Bin Laden, 1993, as quoted in Fadl testimony, *USA* v *Usama bin Laden et al.*, February 6, 2001.
9. Black, op. cit.
10. National Commission, *Staff Statement 5*, March 2004.
11. Ibid.
12. Jason Burke, *Al Qaeda: Casting a Shadow of Terror*, I. B. Tauris, London, 2003, pp. 150–153. Burke argues that the Taliban success did depend on ideology, or more to the point, a lack of a political ideology. Their stress on the purity and power of Islam in all matters effectively replaced political Islam, which he sees as having failed explicitly to gain what it had promised in Afghanistan and the rest of the world.
13. An abridged version of Osama Bin Laden's "Declaration of War Against the Americans" can be found in Appendix C of this book.
14. Philippine National Police investigative files. Also quoted in *USA* v *Yousef et al.* There is some dispute as to the word "Bojinka." Here, it is used as if it were the name of a group that Khalid Sheikh represents. It has generally been applied to the airline bombing plan, not the people, and has been said to mean "explosion" in Serbo-Croatian. The word, however, appears in no Serbo-Croatian dictionaries and native speakers don't recognize it. More likely, it is a slang term Arab fighters used in the Balkans and brought home with them.
15. Neil Herman, interview by Josh Meyer for the *Los Angeles Times*, November 2002.
16. Friends of Abed Sheikh Mohamed, interviews by author, Doha, May 2002.
17. American official, interview by author, Washington, D.C., October 2002.
18. Khaled Mahmoud, interview by author, Doha, May 2002.
19. American official, interview, op. cit.
20. Ibid.
21. Richard A. Clarke, *Against All Enemies*, Free Press, New York, 2004, p. 152.
22. Qatar foreign ministry officials, quoted by FBI Director Louis Freeh in corre-

spondence from Freeh to the Qatari foreign minister, February 1996, author copy.

23. All the participants in the meeting who consented to interviews agreed that Berger pushed to find a way to do it. In general, Berger was quite bullish about renditions.

24. Jamie Gorelick, telephone interview by author, April 2004.

25. Meeting attendee, interview by Josh Meyer for the *Los Angeles Times,* November 2002.

26. Freeh correspondence, op. cit.

27. Meyer, op. cit.

28. Retired CIA officer Robert Baer has told the most colorful version of this story, attributing it to a former Doha police chief. He mentions it in passing in *See No Evil: The True Story of a Ground Soldier in the CIA's War on Terrorism,* Three Rivers Press, New York, 2002, p. 270, but has given more elaborate accounts in numerous interviews.

29. Rodolfo Mendoza Jr., "After Debriefing Report," January 20, 1995.

30. National Commission, *Staff Statement 16,* June 16, 2004. Several news organizations earlier had reported versions of how the hijacking plan came to Al Qaeda. Among the first admittedly speculative accounts was my own, "The Plot," *Los Angeles Times,* September 10, 2002. A year later reports began to appear with far more authoritative reporting on this aspect of the plot. The first was Susan Schmidt and Ellen Nakashima, "Moussaoui Said Not to Be Part of 9/11 Plot," *Washington Post,* March 28, 2003. John Solomon, "9-11 Attack Was to Involve 10 Planes," *Associated Press,* September 22, 2003, advanced the story significantly. The Congressional Joint Inquiry Committee *Final Report,* December, 2002, p. 30, contained the first official confirmation of the sequence of events bringing Khalid Sheikh Mohammed and the airliner plot to Al Qaeda. The National Commission on Terrorist Attacks Upon the United States elaborated on this in much greater detail. None of this later reporting, however, addresses obvious contradictions and inconsistencies in the interrogation reports attributed to Ramzi bin al-Shihb and Khalid Sheikh Mohammed. If the reports are to be believed, bin al-Shibh has given several different versions of events, which, of course, casts doubt on all of them. Additionally, reports attributed to bin al-Shibh are in obvious and frequent disagreement with known facts, such as when people were in certain locations.

31. Indictment in *USA* v *Benevolence International Foundation,* U.S. District Court, Northern District of Illinois, April 29, 2002, p. 15.

32. As with many historic events, Khalid Sheikh's plan in retrospect seemed to have an aura of inevitability, but in fact it is nearer to accident. It came into existence

as the result of three separate, largely unrelated sets of circumstances: Osama Bin Laden being chased out of Sudan, Khalid Sheikh Mohammed being chased out of Qatar, and the Taliban triumph in Afghanistan. Any one of these events by itself might not have had much import beyond whatever its local effects, good and bad. Individually, they presented problems that were potentially significant, but to that point contained. United and hidden from sight behind a veil of ignorance in an area of the world nobody beyond the region cared much about anymore, they would break the bounds of containment. In Afghanistan in the late 1990s, they merged, and in doing so they fundamentally altered the future of Al Qaeda and, thereafter, the global politics of the twenty-first century.

Book Three: The Plot

CHAPTER 1: THE NEW RECRUITS

1. Mounir el-Motassadeq, testifying in his own defense at first trial, Hamburg, October 2002.
2. BKA files, author copy.
3. Camp trainee, interview by author, UAE, August 2003.
4. Shadi Abdullah, testimony in first trial of Mounir el-Motassadeq, Hamburg, November 2002.
5. Omar's nom de guerre is from testimony in the Motassadeq trial; the names for Amir, Jarrah, and Shehhi are from BKA and FBI files and were first reported by Yosri Fouda, on the Al Jazeera television program, "Top Secret: The Road to September 11," September 11, 2002. It's quite likely that Amir, Omar, and Shehhi had taken these names at earlier visits to Afghanistan. In their book, *Masterminds of Terror: The Truth Behind the Most Devastating Attack The World Has Ever Seen,* (Arcade Books, London, 2003), Fouda and Nick Fielding state unequivocally that Amir and Omar were in the camps in 1998. They quote someone they identify as Abu Bakr, "an al Qaeda intermediary," whatever that is, as the basis for the assertion. I agree they were probably in the camps then, too, but *Masterminds* is so rife with casual mistakes that it can't be relied upon. Fouda, to his credit, scored one of the great modern journalistic scoops in the spring of 2002 when he interviewed Omar and Khalid Sheikh Mohammed at a hidden location in Karachi. There was much grousing among other journalists that all Fouda did was answer his telephone and go where he was told, but I'm quite certain anybody else who received that phone call would have quite rightly claimed the exclusivity of their prize reporting. This includes me. Fouda videotaped the interview but was not allowed to keep the tape. Instead, he was later sent audio tapes allegedly made from his originals with only Omar's voice, and that one

doctored to prevent voice-matching. Khalid Sheikh was erased from the tape. In any event, it was a story that Fouda got at some risk to himself and contributed significant information and is to be congratulated. On the other hand, the book, for which the interviews form the foundation, is regrettably and deeply flawed.

6. Interrogations and testimonies of various jihadis over a long period of time indicate that this routine stayed remarkably static for years. Of course, the same could be said for almost all basic military training.

7. Congressional Joint Inquiry Committee, *Report*, pp. 185–187. This is by inference only. There is no direct evidence of the Hamburg men being summoned. The report states that Mohammed Haydar Zammar, the man who is suspected of recruiting the pilots, was in direct contact with senior Al Qaeda officials in the months immediately before the men came to Afghanistan.

8. Fouda, *Masterminds of Terror*, and the National Commission, *Final Report*, July 22, 2004, p. 166.

9. Rabbani, interview, op. cit.

10. Jamal Ahmed al-Fadl, testimony in *USA* v *Usama bin Laden et al.*, February 6, 2002.

11. If you didn't mind the fact that he sometimes refused to answer questions and instead simply spoke what was on his mind, he was a good interview. Many of these interviews or transcripts thereof are available online. Examples can be found at: http://news.findlaw.com/cnn/docs/binladen/binladenintvw-cnn.pdf and: http://abcnews.go.com/sections/world/ DailyNews/ transcript_binladen1_981228.html.

12. U.S. intelligence estimate, June 2002.

13. Stephen Cohen, quoted in "The Cost of Being Osama Bin Laden," *Forbes* Magazine, September 14, 2001.

14. National Commission, *Final Report*, July 22, 2004, p. 156.

15. The Saudi government effectively disinherited Bin Laden in 1994 when it ordered Bin Laden's share of the family company sold and froze the proceeds, according to the National Commission on Terrorist Attacks Upon the United States, *Terror Financing Monograph*, August 20, 2004.

16. The National Commission estimated that Bin Laden's net worth was never more than $20 million; see *Staff Statement 15*, June 16, 2004.

17. The financing of Al Qaeda is beyond the scope of this book. The fact is the September 11 attacks didn't cost that much money—somewhere in the neighborhood of $250,000 to $500,000. They could have been funded out of relative pocket change. That said, substantial money flows were necessary to enable construction of the terrorist infrastructure that supported the attacks. Al Qaeda's budget has been estimated at $30 million per year by the CIA, according to the

National Commission. As with many other aspects of Al Qaeda, the organization took advantage of traditional Muslim practices in building this financial base. Charitable giving, *zakat*, as it is called, is one of the pillars of Islam, and all Muslims are expected to give at least 10 percent of their income annually to charity. *Zakat* almost certainly constitutes the largest organized charitable giving on Earth. Combine that religious-inspired inclination with the avowal for a decade that the war in Afghanistan and subsequent wars in Bosnia, Kosovo, and Chechnya were holy wars against infidels, and it is little wonder world governments have had difficulty stopping the flow of funds.

18. Herman, interview by Josh Meyer, op. cit.

19. Interviews with European and American investigators, spring 2002.

20. Ibid.

21. National Commission, *Final Report*, July 22, 2004, p. 153.

22. Senior Justice Department official, interview by Josh Meyer for the *Los Angeles Times*, Washington, November 2002.

23. George Tenet, "Written Statement of Director of Central Intelligence George Tenet Before the Congressional Joint Inquiry Committee," October 17, 2002.

24. Federation of American Sciences, online archives. Accessed at http://www.fas.org/irp/world/para/docs/980223-fatwa.htm.

25. National Commission, *Staff Statement 16*, June 16, 2004.

26. Ibid., and John Solomon, "9-11 Attack Was to Involve 10 Planes," *Associated Press*, September 22, 2003.

27. National Commission, *Staff Statement 16*.

28. Solomon. "9-11 Attack."

29. Nickels, interrogations, op. cit.

30. Bruhn, interview by author, Hamburg, April 2002.

31. Fateh, interview, op. cit.

32. Jarrah's relationship with Aysel has led to the conventional depiction of him as someone torn between the two worlds and thus one whose beliefs were less extreme and devotion more questionable. It could, as well, be the opposite—that his beliefs were more extreme and his devotion greater so as to overcome his other commitments.

33. National Commission, *Final Report*, July 22, 2004, p. 166. The first meeting with Bin Laden did not include Omar, who had yet to arrive in Kandahar. He met separately with Bin Laden after his arrival and, as had the others, swore allegiance.

34. The video was later among a group found in Al Qaeda leader Mohammed Atef's Kabul residence during the U.S.-Afghan war. Others in a numbered series found with it were date-stamped March 10. Omar used a credit card in the Netherlands on March 13. Some people (see note 43, below) believe he may have gone from

Karachi to Kuala Lumpur to attend the Hambali meeting there, but there is no definitive proof of his location anywhere between January 3 and March 13, according to BKA files, and interviews with German and American investigators, and Malaysian security officials, 2002.

35. International Crisis Group, "Jemaah Islamiyah in South East Asia"; see also Mendoza, "Philippine Jihad Inc."

36. Mohamad Sobri, interview by author, Sungai Manggis, June 2002.

37. Ibid.

38. International Crisis Group, "Jemaah Islamiyah in South East Asia," p. 4, places Hambali at Abdur Rasul Sayyaf's training camp in 1987.

39. Zachary Abuza, quoted in Don Greenlees, "Terrorism—Still a Force to Be Feared," *Far Eastern Economic Review*, Jakarta, January 22, 2004.

40. National Commission, *Final Report*, July 22, 2004, p. 150.

41. Any other purpose of the meeting has never been firmly established. Several of the men at it could just as easily have met in Afghanistan or elsewhere. It's possible the presence of all the men there at the same time was as much a matter of coincidence as collusion. In retrospect, what was important about the meeting was who was there, not what they were doing, and the broadly shared sense that the 9/11 plot could have been sidetracked there before it ever really got underway.

42. Congressional Joint Inquiry Committee, *Report*, p. 145.

43. Malaysian security officials and American and German investigators, interviews by author, Kuala Lumpur, Washington, Berlin, Spring-Summer 2002. There is some dispute about what the photographic record actually shows. Some people who have seen it claim that in addition to the people mentioned, there are also photographs of Ramzi bin al-Shibh (Omar) at the condo. Some people have identified him from the photographs; others say it is not clear it is he. Staff reports of the Congressional Joint Inquiry Committee and the National Commission on Terrorist Attacks Upon the United States deal with the Malaysia meeting in enlightening detail. Neither mentions Shibh's presence. There is no travel record for him, but alone among the Hamburg group he was known to use false-identity documents. Shibh did go to Malaysia eighteen months later, in June 2001.

44. National Commission, *Final Report*, July 22, 2004, p. 156.

45. American officials say they failed to discern that Mihdhar and Hazmi entered the United States. Why this is so is an enduring mystery. Mihdhar's passport, of which the CIA had a copy, contained an extended-stay visa for the United States, a fact intelligence officers confirmed with consular officials who issued the visa in Saudi Arabia. Knowing that, and thinking that Mihdhar and Hazmi were Al Qaeda operatives and "something nefarious might be afoot," as one CIA officer

described it, the lapse in tracking them is inexplicable. It wasn't that they thought the meeting was insignificant: Eight CIA stations around the globe were involved in tracking its participants; top national security officials were briefed on it, including George Tenet, director of central intelligence, and Sandy Berger, national security adviser. For the event to have that much importance one week and be dropped almost entirely the next speaks both to the very high level of suspected terrorist activity that was being tracked, and to fundamental flaws in the intelligence services. Beyond the CIA's failure, the National Security Agency, which intercepts signals such as telephone calls, had identified Mihdhar and both Hazmi brothers many months before the meeting as Al Qaeda agents. They also identified them en route to the meeting, yet they never informed any other agencies of this and they, too, failed to notice that the men entered the United States.

CHAPTER 2: PREPARATIONS

1. This account of Aysel Sengün's activities is drawn primarily from BKA investigative files, which include Sengün's correspondence and extensive interviews with her and with others having direct knowledge of the events. It also relies to a more limited extent on public testimony in the Hamburg trial of Mounir el-Motassadeq.

2. Motassadeq, testifying at his first trial, Hamburg, October 23, 2002.

3. Senior security official, UAE, interview by author, Abu Dhabi, July 2003. The Emiratis say that they initially stopped Jarrah because his name was on a watch list provided to them by the United States. American officials say this is untrue. The United States, however, has acknowledged in internal documents and in communications with German investigators that the Emiratis did contact them about Jarrah. They decline to say what they told the Emiratis. The Americans told the Germans, according to FBI documents obtained by the author, the Jarrah interview was not substantive. Merely routine, they said. The Emiratis say this is utterly untrue. The interview lasted four hours, according to their records, and the Americans were informed of it while it was occurring. It is possible that the facts of the event were victims of the dysfunctional relationship among different American agencies. The Emiratis won't say what Americans they contacted, but it would be a good guess that they called the CIA station at the U.S. embassy in Abu Dhabi and that the information either died there or was passed back to Langley and in either case not shared with the FBI.
At this point, lacking further documentation from either side, it is impossible to say which version is correct. It is worth noting, however, that when the initial reports of the Jarrah interview were made by Jane Corbin for the BBC in 2001 (and later repeated in her book, *Al Qaeda: The Terror Network That Threatens the*

World, Nation Books, New York, 2002), citing UAE sources, the Americans publicly denied they had ever been informed of it. As it happened, Corbin had the wrong date for the event, so the American services might have been technically correct in denying any knowledge of it. They later repeated that denial several times when other reports repeated the inaccurate date. As of this writing Americans have yet to correct the record, although, as noted above, the FBI has acknowledged to its German counterpart the stop did occur and they were informed.

4. Sengün interrogations quoted in BKA investigative files, author copy.

5. Correspondence contained in BKA investigative files, author copy.

6. Congressional Joint Inquiry Committee, *Report,* p. 173.

7. A friend of Hazmi's said Hazmi gave him this account; interview by Gil Reza for the *Los Angeles Times,* October 2001. Reza's reporting is the foundation for much of the material in this section. Much of that reporting was subsequently confirmed first by the Congressional Joint Inquiry Committee and then by the National Commission on Terrorist Attacks Upon the United States.

8. Reza interviews, op. cit., September-October 2001.

9. Bayoumi visited the Saudi consulate the very day he first met Hazmi and Mihdhar.

10. Congressional Joint Inquiry Committee, *Final Report,* op. cit.

11. Ibid, pp. 131–132. The Congressional Joint Inquiry Committee gave this summary of their backgrounds: "Al-Hazmi first traveled to Afghanistan in 1993 as a teenager and came into contact with a key al-Qa'ida facilitator in Saudi Arabia in 1994. In 1995, al-Hazmi and al-Mihdhar traveled to Bosnia to fight with other Muslims against the Serbs. Al-Hazmi probably came into contact with Al-Qa'ida leader Abu Zubaydah when Zubaydah visited Saudi Arabia in 1996 to convince young Saudis to attend al-Qa'ida camps in Afghanistan. Sometime before 1998, al-Hazmi returned to Afghanistan and swore loyalty to Bin Ladin. He fought against the Northern Alliance, possibly with his brother Salem, another of the hijackers, and returned to Saudi Arabia in early 1999. . . . Al-Mihdhar's first trip to the Afghanistan training camps was in early 1996. In 1998, al-Mihdhar traveled to Afghanistan and swore allegiance to Bin Ladin. In April 1999, Nawaf al-Hazmi, Salem al-Hazmi, and Khalid al-Mihdhar obtained visas through the U.S. Consulate in Jeddah, Saudi Arabia. Al-Mihdhar and Nawaf al-Hazmi then traveled to Afghanistan and 'participated in special training. . . .'"

12. As incredible as it seems, Shaikh had in fact been a long-time paid informer for the FBI and testified, albeit without being publicly identified, before the Congressional Joint Inquiry Committee. He told the committee he had no idea Hazmi and Mihdhar were potential terrorists and so never discussed them with the FBI. Neither did the FBI ever ask him to be on the watch for men taking flying lessons.

13. Reza interviews, op. cit.

14. The money actually came from one of a pair of men who were the main conduits of funding for the September 11 attacks, Ali Abdul Aziz Ali; Ali is Khalid Sheikh Mohammed's nephew.

15. Friend of Hazmi, interview by Gil Reza for the *Los Angeles Times*, San Diego, December 2001.

16. Abdussattar Shaikh, interview by Gil Reza for the *Los Angeles Times*, San Diego, September 2001.

17. Ibid.

18. The entire San Diego aspect of the September 11 plot begs explanation. Mihdhar and Hazmi's associations both before and after arriving in the United States make it seem extremely unlikely authorities would not become more interested in them more quickly and followed that interest more diligently. The facts that they were known to be Al Qaeda operatives, that they were helped to settle in San Diego by someone working at least informally with the Saudi government, that they just happened to live with an FBI informer combine to make one wonder how they were not caught.

    In addition to those events already noted, Hazmi placed telephone calls from a mobile phone registered to him in San Diego to the number in Yemen that was under surveillance by U.S. intelligence services. This was the same telephone number, belonging to the Hada clan, that was the source of the original intercepted information about Mihdhar and Hazmi going to the meeting in Kuala Lumpur. The United States has not said so, but monitoring of that Yemen telephone presumably continued, so the calls Hazmi made from the United States would routinely have been intercepted. One would think that would have set off some sort of alarm in the intelligence community, but apparently it did not. The series of missed opportunities indicates they were not isolated failures but evidence of systematic deficiencies in the intelligence community's ability to both collect and, especially, analyze information.

19. BKA files, author copy, July 2002.

20. E-mail retrieved and quoted in BKA files.

21. Ibid. The friend replied in part: "I should be angry at you, but of course I can't. I don't dare. But I was really worried about you. I called you in Egypt but they don't know your number, and I called you in Germany and they don't know your number. WHERE R U YA MAN."

    About the visa requirement: Whether Amir learned the correct process or not, he didn't follow it. He was correct in wondering if a student visa was supposed to be obtained before entering the country to study. It is. Instead, he and Shehhi both traveled to the United States on visitors' visas, and didn't apply for student visas until after they had started school, which was technically illegal.

22. National Commission, *Staff Statement 1*, January 26, 2004, p. 5.

23. BKA files. In Germany, this was not a casual event. In order to by placed on such lists, intelligence agencies had to go to great lengths to demonstrate to the Bundestag, the German parliament, that the person under question was of potential danger to the state. Being placed on the list indicated that Motassadeq had been under investigation for some time. In that he was an integral part of the group that included Amir, Omar, and Shehhi, this at the least implies that they were being watched, too.

24. Motassadeq, testifying at his first trial, op. cit.

25. These events were all revealed at Motassadeq's trial and elaborated upon in prosecutor's investigative files, author copies.

26. Ramzi bin al-Shibh, interview with Yosri Fouda for Al Jazeera, Karachi, April 2002; translation by Naouar Bioud, author copy. There has been a persistent story reported that Amir made two trips to Prague before coming to the United States. The source of that appears to be confusion on the part of Czech Republic passport control officials. Another man with a similar name traveled to Prague from Saudi Arabia on the same day and was denied entrance because he lacked the correct visa. That man then went to Germany, where he obtained the visa. Amir made a simple, unremarkable transit through Prague.

27. This name confusion has given rise to the notion that Atta used aliases either in Germany or the United States. That does not appear to be the case. His full name was Mohamed Mohamed el-Amir Awad el-Sayed Atta. On different rental forms, credit agreements, visas, etc., he was variously referred to as Atta, el-Amir, and, occasionally, Sayed. Similarly, Shehhi's original UAE passport was issued with the last name Lekrab, which again was a name he never used. The UAE changed its naming conventions for public documents in the late 1990s, and all identification documents from then on used the new convention. In Shehhi's case, that meant that Lekrab was dropped and the name his family used, Shehhi, was the name on all public documents from then on.

28. FBI investigative records, author copy.

29. Airman Flight School had been a favorite of Al Qaeda trainees for years, a fact well known to American security services, who assumed the trainees would be employed flying Bin Laden's personal aircraft and Al Qaeda transport planes.

30. Atta and Shehhi banking records, author copy.

31. Carol J. Williams, and John-Thor Dahlburg, and H. G. Reza, "Mainly, They Just Waited," *Los Angeles Times*, September 27, 2001.

32. National Commission, *Final Report*, July 22, 2004, p. 214.

33. FBI investigative records, author copy. As further evidence of the strength of the

family ties, Ali's sister, Latifa, was married to the original World Trade Center bomber, Ramzi Yousef.

34. Atta and Shehhi banking records, op. cit.

35. Although both passed, Atta scored significantly better on the exams. The two of them were considered sufficiently skilled as pilots that the owner of the school later offered them jobs as pilots on a new airline he was planning to start.

36. FBI files, author copy.

37. BKA files, op. cit.

38. Ibid.

39. Sengün, interrogations, op. cit.

40. E-mail quoted in BKA files, author copy.

41. Huffman Aviation flight logs, author copies.

42. The tape was later found in the Kabul home of Mohammed Atef, Bin Laden's military commander, who died under American bombing attacks late in 2001.

43. This account is drawn from BKA interrogations with the woman.

44. FBI records, op. cit.

45. Ibid.

46. Motassadeq interrogations, BKA files, op. cit.

47. Indictment in *USA* v *Zacarias Moussaoui*, Eastern District of Virginia, July 2002.

48. Mark Fineman, "A Case of Where, Not What: Prosecutors Say Zacarias Moussaoui's Travels—from an Afghan Terrorist Training Camp to an Oklahoma Flight School—Are Key to His Indictment in the September 11 Attacks," *Los Angeles Times*, March 30, 2002.

49. This would typically have been Abu Zubaydah (né Mohammed Hussein Zein-al-Abideen), the Palestinian who oversaw the organization's camps, inviting the most promising candidates for more advanced training.

50. Susan Schmidt and Ellen Nakashima, "Moussaoui Said Not to Be Part of 9/11 Plot," *Washington Post*, March 28, 2003.

51. Ibid.

52. BKA files, op. cit.

53. Schmidt and Nakashima, "Moussaoui . . ."

54. National Commission, *Staff Statement 16*, June 16, 2004.

55. Susan Khalil, telephone interview by Patrick McDonnell for the *Los Angeles Times*, June 2002.

56. Williams, Dahlburg, and Reza, "Mainly, They Just Waited."

57. Rich Connell, "Response to Terror Tricky Pursuit," *Los Angeles Times*, October 21, 2001.

58. Charles Sennott, "Why Bin Laden Plot Relied on Saudi Hijackers," *Boston Globe*, October 3, 2002.

59. Abdussattar Shaikh, interview by Gil Reza for the *Los Angeles Times*, October 2001.

60. Pakistani recruit, interview by author, Abu Dhabi, July 2003. The recruit said Mohammed gave his name simply as Khalid Sheikh. He provided a detailed physical description of him and later identified him from a selection of photographs.

61. A radical mosque in Hamburg, Germany, was controlled by people who called themselves Tabligh; they financed travel for several people to Afghanistan training camps. Additionally, the Hamburg pilots were recruited for Al Qaeda by a Tabligh.

62. Pakistani recruit, interview by author, op. cit.

63. Senior Gulf intelligence officer, interview by author, August 2003, UAE.

64. Shadi Abdallah, testimony, op. cit., November 27, 2002.

65. Senior Italian investigator, telephone interview by Sebastian Rotella for the *Los Angeles Times*, July 2002.

66. BKA records, op. cit.

67. National Commission, *Final Report*, July 22, 2004, p. 191.

68. *USA* v *Agus Budiman*, Eastern District of Virginia, November 16, 2001.

CHAPTER 3: THE LAST YEAR

1. Atta and Shehhi came close to not even getting back into the country. Both had initially come to the United States on six-month visitor's visas, which had expired by the time they made these second entries. Both had already applied for new student visas, but had yet to receive them. Technically, neither should have been let into the country, and both were stopped and questioned by immigration agents, who noted the expired visas. They were sent to secondary inspection areas for more detailed questioning. Both told immigration agents they were flight students and were awaiting approval of their requests to change status. Both were admitted again and given new tourist visas. Shehhi's was again for six months. Atta's inexplicably was for the odd period of eight months. It would not expire until September 8. The visa errors, then, didn't hurt the men. In fact, in Atta's case, it actually would have helped him stay in the United States longer. (The eight-month period was later shortened to the standard six when Atta took another man to an Immigration and Naturalization Service office and tried to get him an eight-month visa like his. The officer who heard the request declined to give the other man the extra two months and took Atta's away from him. In a very round-about way, the system worked.)

What is most notable about the visa foul-up is not that the government made an error, although clearly it did, but the carelessness of the plotters. It illustrates again the degree to which Al Qaeda was not a slick, professional outfit that didn't get caught because it didn't make mistakes. It made mistakes all the time. It didn't get caught because the government with which it was dealing made more of them.

2. Credit card records quoted in FBI investigative files, author copy.

3. The pair returned to Virginia Beach two months later, in early April. This time they withdrew twice as much cash—$8,000—and closed the Mailbox account. There is no indication whatsoever what this was about, but Hazmi and Hanjour arrived in Virginia at about that same time. The cash could have been for them, especially considering that they had no other apparent source of funding. The earlier trip could have been a dry run to insure they could access their account and make large deposits away from their Florida base.

4. FBI investigative records, author copy.

5. National Commission, *Staff Statement 16*, June 16, 2004.

6. Sengün, interrogations, op. cit.

7. BKA investigative files, author copy.

8. Quoted in BKA files, ibid.

9. Security official, UAE, interview by author, Abu Dhabi, July 2003.

10. Motassadeq, testifying at his first trial, in Hamburg, October 2002.

11. U.S. authorities were so desperate to find out what they did and who they met with that they later arrested a man on the apparent grounds he was their waiter at one of these restaurants. Dan Christensen has written extensively about this for the *Miami Daily Business Review*. See, for example, "Low Burden of Proof. Coincidence, Uncorroborated Report Enough to Get Arab Waiter in Delray Beach Detained Five Months," March 14, 2003.

12. Interview by John Thor-Dahlburg for the *Los Angeles Times*, Delray Beach, Florida, September 2001.

13. National Commission, *Final Report*, July 22, 2004, p. 252.

14. Why he needed to meet Atta's father is uncertain. The father told investigators Shehhi came to pick up Atta's international driver's license, but further investigation indicated Atta had the license with him already.

15. *USA* v *Zacarias Moussaoui*, Eastern District of Virginia, July 2002.

16. Khalid Sheikh Mohammed and Ramzi bin al-Shibh have given conflicting accounts of whether Moussaoui was ever seriously intended as a 9/11 pilot. Shibh indicates he was, Mohammed that he was not and that he was intended instead to take part in a second wave of attacks that was never launched. See National Commission on Terrorist Attacks Upon the United States, *Staff Statement 16*, June 16, 2004.

17. Ibid.

18. Ibid.

19. Waheed Hamza Hashim, interview by author, Jeddah, August 2003.

20. Yosri Fouda, "Top Secret: The Road to September 11," Al Jazeera, September 11, 2002.

21. John Solomon, "9-11 Attack Was to Involve 10 Planes," Associated Press, September 22, 2003. Mihdhar's absence from the United States for the entire time during which these men were brought in to the plot suggests he might have played some role in assisting them. Khalid Sheikh Mohammed, according to the National Commission, *Staff Statement 16*, wanted to remove Mihdhar from the plot entirely because he had left California in summer 2000, without permission. Mihdhar told friends in San Diego he had to return to his family. Bin Laden later interceded on Mihdhar's behalf and insisted he be part of the 9/11 attacks, although there is no evidence to support the idea that he was given any additional role. According to interrogation transcripts of a captured Al Qaeda operative, he was seen with three of the younger recruits at a training camp near Kandahar in March 2000. He returned to the United States after more than a year's absence just five days after the last of the other men arrived.

22. There has been much speculation about Bin Laden's decision to include so many Saudis. Most analysts—including most prominently those within the Saudi government—have concluded that it was a means for Bin Laden to attack the Saudi monarchy, and thereby embarrass and weaken it.

23. National Commission, *Staff Statement 1*, January 2004. See also Joel Mowbray, "Visas for Terrorists: They Were Ill-Prepared. They Were Laughable. They Were Approved," *National Review*, October 28, 2002, for an earlier examination of the visa processes.

24. Al Jazeera broadcast, September 9, 2002.

25. Fouda, "Top Secret."op. cit.

26. The nephew, Ali Abdul Aziz Ali, had been given the money, about $10,000 apiece, by Khalid Sheikh. Where it originated is still not known. Although the banks in Dubai were largely unregulated and funds were difficult to track, Ali kept the money in a laundry bag in his flat in Dubai, according to post-September 11 intelligence cited by the National Commission. Ali had a day job at a computer wholesaler and complained to Khalid Sheikh that he didn't have time to do both that job and the jihad work. Khalid dispatched a member of Al Qaeda's media committee to Dubai to assist him.

27. BKA files, op. cit.

28. Sengün said Jarrah had a longstanding interest in martial arts and had taken lessons even as a kid in Beirut.

29. Belfas, quoted in BKA investigative files.

30. The reporting on the Prague meeting was based on a single retrospective eye-witness identification by a Czech informant who was following the man Atta was alleged to have met on April 9 in Prague. There is nothing else to support such a meeting, although the U.S. security services say they spent thousands of hours investigating it. What is known is that Atta was in Virginia on April 4 and in Florida on April 11. In between, his cell phone was used in Florida on April 6, 9, 10, and 11, according to FBI records quoted by the National Commission on Terrorist Attacks Upon the United States. Additionally, the man Atta was supposed to have met was taken into U.S. custody during the Iraq war and denied the meeting occurred. It also turns out that he was not in Prague at the time of the alleged meeting. The meeting became an ideological touchstone in the later U.S. debate over alleged connections between Al Qaeda and Iraq. Those searching for evidence of such a connection, most prominently including Vice President Dick Cheney, declined to drop the matter despite the lack of evidence. As Defense Secretary Donald Rumsfeld liked to say about other searches, the absence of evidence didn't necessarily imply the evidence of absence and it is at least possible such a meeting occurred. But you would have to wonder why Al Qaeda would take such a risk. What would be the purpose of sending someone they had been patiently training for more than a year on a whirlwind and risky undercover mission for a meeting with a known spy. If they wanted to meet with Iraqis, wouldn't it be much simpler to do it somewhere safer, Afghanistan, Pakistan, even Iraq? And what would Iraq have been able to contribute to the September 11 enterprise? Money? They already had plenty to achieve what they needed. If September 11 marked the opening salvo of a war, it was one of the least expensive battles ever fought. Expertise? Atta, not the spy, was the trained pilot. Al Qaeda made all sorts of mistakes and many demonstrably silly decisions; this seems very unlikely to have been one of them.

31. FBI telephone records, author copy. The meeting was originally scheduled to take place in Malaysia, and Omar flew there using a false Saudi identity, but Atta for unknown reasons was unable to make it and the location was changed to Spain, which was more easily and quickly accessible from the eastern United States, according to the National Commission on Terrorist Attacks Upon the United States. *9/11 and Terrorist Travel*, August 20, 2004, p. 5.

32. Sebastian Rotella, "Inquiry into Madrid Cell Provides a View of Al Qaeda's Wide Reach," *Los Angeles Times*, September 1, 2002. Rotella's reporting on the Spanish Al Qaeda connections has been unsurpassed and is the basis for much of this account.

33. Neighbors, interviewed by author, Madrid, May 2002.

34. Rotella, op.cit.

35. Rian Tatari Bakhri, author interview, Madrid, May 2002.

36. Sebastian Rotella, "Clues Indicate That a Support Network in Spain and Germany Figured in September 11 Plot. Top Suspects Are a Clique of Syrian Immigrants," *Los Angeles Times,* January 14, 2003.

37. Jose Maria Irujo, "September 11 Attack Plot Finalized in Spain," *El Pais,* March 2, 2004.

38. National Commission, *Staff Statement 16,* June 16, 2004.

39. Ibid.

40. Ibid.

41. Ibid.

42. BKA files.

43. Mark Fineman, "A Case of Where, Not What."

44. Moussaioui, as quoted in FBI interrogations, October, 2001, author copy.

45. Here, for example, are the titles of a series of reports from the CIA to the rest of the government from the time period:

"Sunni Terrorist Threat Growing," February 6, 2001

"Bin Ladin Planning Multiple Operations," April 20, 2001

"Bin Ladin Public Profile May Presage Attack," May 3, 2001

"Terrorist Groups Said Cooperating on U.S. Hostage Plot," May 23, 2001

"Bin Ladin Network's Plans Advancing," May 26, 2001

"Bin Ladin Attacks May Be Imminent," June 23, 2001

"Bin Ladin and Associates Making Near-Term Threats," June 25, 2001

"Bin Ladin Planning High-Profile Attacks," June 30, 2001

"Bin Ladin Threats Are Real," June 30, 2001

"Planning for Bin Ladin Attacks Continues, Despite Delays," July 2, 2001

"Bin Ladin Plans Delayed but Not Abandoned," July 13, 2001

"One Bin Ladin Operation Delayed, Others Ongoing," July 25, 2001

"Bin Ladin Determined to Strike in U.S.," August. 7, 2001

46. For example: The National Security Agency was charged with worldwide collection of ELINT, or electronic intelligence, and SIGNINT, signals intelligence (telephone conversations). That haul was considerable. NSA estimated, to choose but one target, that worldwide international telephone calls totaled 180 billion minutes in 2003. NSA routinely took in so much material that as a matter of policy it didn't even try to sort through what it hauled in unless it was specifically asked to do so. It saw its mission as a provider to other intelligence consumers; if no one came to the drive-through window and placed an order, they didn't bother to cook.

47. Congressional Joint Inquiry Committee, *Report,* p. 171. The CIA had at least one informant with access to Al Qaeda camps.

48. Thomas Pickard, deputy director of the FBI, testimony before the National Commission, April 13, 2004.

49. Ibid.

50. In the end, the Moussaoui arrest caused more upset than action. People were briefed; no one did anything. It might have had more effect among the hijack crews. Omar was in constant communication with Atta, Jarrah, and Shehhi. He talked with them every two or three days. Presumably, he did the same with Moussaoui. Not hearing from him at all after a certain point would have set off alarms up and down the network. If there was any consideration given to delaying the plot at that point, Moussaoui's disappearance would have been a powerful argument against it.

51. The search was at best leisurely. According to the National Commission, *Staff Statement 10*: "The search was assigned to one FBI agent for whom this was his very first counterterrorism lead. By the terms of the lead, he was given 30 days to open an intelligence case and make some unspecified efforts to locate Mihdhar. He started the process a week later. He checked local New York indices for criminal record and driver's license information and checked the hotel listed on Mihdhar's U.S. entry form. On September 11 the agent sent a lead to Los Angeles based on the fact that Mihdhar had initially arrived in Los Angeles in January 2000."

    Numerous news organizations have reported that one way they might have gotten quicker results into the whereabouts of the pair was to check telephone directories. Hazmi was listed in the San Diego phone book for two years. They might also have asked their own informants, particularly those in California, where the pair had first entered the country. Recall that Hazmi and Mihdhar rented a room in the home of one of the FBI's best counterterrorism informants in San Diego. One wonders, though, what it means to be among the agency's best informants if he neglected to tell anybody about the two Saudis living in his back bedroom.

52. National Commission, *Staff Statement 1*, January 26, 2004, p. 5.

53. Ibid.

54. INS Officer Jose Melendez-Perez, testifying at the National Commission, public hearing, January 26, 2004. Melendez-Perez said there was normally extraordinary deference given to Saudi citizens, but this man's attitude was such that the deference was overwhelmed. Even at that he was warned by fellow workers that he could get in trouble for hassling a Saudi. Saudis were frequent visitors to Disney World, the local theme park, and INS agents knew that if they displeased the visitors they were apt to hear about it. This is yet one more piece of infor-

mation illustrating the degree to which economics trumped security concerns. Nonetheless, Melendez-Perez recommended the man not be admitted and be sent back home on the first available flight; his supervisors backed his decision without hesitation. The information that Atta was in the terminal comes from FBI telephone records quoted by the commission staff.

55. National Commission, *Staff Statement 16*, June 16, 2004.

56. Yosri Fouda and Nick Fielding, *Masterminds of Terror*, p. 215.

57. Ramzi bin al-Shibh, interview with Yosri Fouda, op. cit. The date is from the transcript. That Omar sent Essabar comes from the National Commission, *Staff Statement 16*, June 16, 2004.

58. Interrogation transcripts of Mohammed Jabarah, a young Canadian-Kuwaiti who had joined Al Qaeda in 2000, been trained in Afghanistan, then selected—largely because of his English skills and Canadian passport—to coordinate a cell in Southeast Asia that was plotting to bomb numerous Western targets throughout the region. It was a measure of Khalid Sheikh Mohammed's tenacity that he was already looking beyond September 11 to the next round of attacks. He and Hambali spent the weeks just prior to September 11 instructing Jabarrah on communications protocols. One means of communication was to be e-mail. Khalid Sheikh told Jabarrah his e-mail addresses were silver_crack@yahoo.com and gold_crack@yahoo.com.

Singaporean authorities disrupted the plot Jabarrah was coordinating, but were unable to capture all of the men in it. The plot later evolved into the horrific Bali nightclub bombing, the ingredients for which were purchased with money Zacarias Moussaoui was to have used for his flight training in Malaysia.

59. Employer, interview by author, Hamburg, May 2002.

60. Arens, interview, op. cit.

61. BKA files, author copy.

62. The initial speculation, which might be correct, was that Atta wanted to reduce the number of hijackers coming through security at Boston's Logan Airport. Two of the four target flights were departing from there, meaning ten hijackers had to board. The notion that one could go to nearby Portland, board a commuter flight to Boston, get inside the security bubble, and connect with the target flight by essentially a different route seemed on the face of it logical. Go to a smaller airport with presumably more lax security and avoid the tougher security at a major international airport. It made sense; the facts, however, didn't fit. Because the commuter flight landed at a different concourse than the target flight would depart Logan from, Atta had to exit his arrival concourse and enter the other, going through security for a second time. So rather than lessening his

exposure to security, he actually doubled it. Additionally, security at Portland turned out to be in a way tougher. It had video cameras at boarding points. Logan did not. The videotape depicting Atta and al-Omari boarding their flight that day is of the commuter flight from Maine, not the targeted American transcontinental flight. Finally, the narrow time connection between the two flights greatly raised the risk Atta and al-Omari could miss their Boston flight entirely. As it happened, they nearly missed their first flight, arriving at the airport just twenty minutes before it took off. They arrived in Boston on time, but for some reason their luggage did not make the flight. Atta's bag contained nearly every important document in his life—his will, the "Last Night" instructions, his high school diploma, his TUHH diploma, his original visa for Germany from Egypt, a power of attorney, employee evaluations, even the certificate from the Goethe Institute in Cairo where he had learned German. If you wanted to leave a roadmap for investigators to follow, the suitcase was a pretty good place to start.

63. BKA files, author copy.
64. Shibh, interview with Yosri Fouda, op. cit.
65. See Appendix B for the complete instructions.

CHAPTER 4: THAT DAY

1. Shibh, interview with Yosri Fouda, op. cit.
2. FBI records, author copy.
3. Ibid.
4. John Raidt, staff member of the National Commission, testimony before the National Commission, January 27, 2004.
5. That they are quite easily able to check passengers' names against the even larger list of their frequent-flyer programs seems not to have occurred to the FAA.
6. Shibh, interview with Yosri Fouda, op. cit.
7. Ibid.
8. Air Force Colonel Alan Scott, testimony before the National Commission, May 23, 2004. The indicated times vary somewhat from FAA time lines for the same events mainly because the NORAD time line was constructed to reflect the times it logged the events.
9. National Commission, *Staff Statement 17*, June 17, 2004.
10. Norman Mineta, transportation secretary, testimony before the National Commission, May 23, 2004.
11. National Commission on Terrorist Attacks Upon the United States, *Staff Statement 17*, June 17, 2004.
12. The hijack pilots had agreed among themselves ahead of time that they would

crash the planes rather than lose control of them to passengers or crew. Atta had planned to crash his aircraft into the streets of New York if he could not hit the World Trade Center, according to the National Commission, *Staff Statement 16*, June 16, 2004.

13. John Skilling, interview by author, Seattle, 1989.

14. Jon Magnusson, telephone interview by author, September 11, 2001.

APPENDIXES

Note that the Appendixes are rendered exactly as they appear in the originals, with spelling and grammatical errors left intact.

Appendix A: Originally published by *Der Spiegel*, Hamburg, Germany; translation by ABC News.

Appendix B: Translation by Capital Communications Group and Imad Musa, for the *New York Times*. September 29, 2001.

Appendix C: Originally published by *Al Quds Al Arabia*, London, August 1996.

Appendix D: Federation of American Sciences. Accessed at http://www.fas.org/irp/world/para/docs/980223-fatwa.htm.

Selected Bibliography

Akbar, M.J. *The Shade of Swords*, Routledge, New York, 2002.

Azzam, Abdullah. *Join the Caravan*, Azzam Publications, London, 2001.

Baer, Robert. *See No Evil: The True Story of a Ground Soldier in the CIA's War on Terrorism*, Three Rivers Press, New York, 2002.

Benjamin, Daniel and Steven Simon. *The Age of Sacred Terror*, Random House, New York, 2002.

Bergen, Peter. *Holy War, Inc.*, Touchstone, New York, 2002.

Burke, Jason. *Al Qaeda: Casting a Shadow of Terror*, Tauris, London, 2003.

Clarke, Richard A. *Against All Enemies*, Free Press, New York, 2004.

Coll, Steve. *Ghost Wars: The Secret History of the CIA, Afghanistan, and Bin Laden, from the Soviet Invasion to September 10, 2001*, Penguin Press, New York, 2004.

Cooley, John K. *Unholy Wars: Afghanistan, America, and International Terrorism*, Pluto Press, London, 2000.

Edwards, David. *Before Taliban*, University of California Press, Berkeley, 2002.

Fouda, Yosri and Nick Fielding. *Masterminds of Terror: The Truth Behind the Most Devastating Terrorist Attack the World Has Ever Seen*, Arcade, London, 2003.

House Permanent Select Committee on Intelligence and the Senate Select Committee on Intelligence. *Report of the Joint Inquiry Committee into the Terrorist Attacks of September 11, 2001*, December 2002.

Lewis, Bernard. *The Crisis of Islam: Holy War and Unholy Terror*, Modern Library, March 2003.

National Commission on Terrorist Attacks Upon the United States. *Staff Reports*, 2003–2004, and, *Final Report*, 2004.

Rashid, Ahmed. *Taliban*, Pan Books, London, 2001.

Reeve, Simon. *The New Jackals*, Northeastern University Press, Boston, 1999.

Ressa, Maria. *Seeds of Terror*, Free Press, New York, 2004.

Riebling, Mark. *Wedge: From Pearl Harbor to 9/11*, Touchstone, New York, 2002.

Rubin, Barnett R. *The Fragmentation of Afghanistan*, 2d ed., Yale University Press, New Haven, 2002.

Rubin, Barry and Judith Colp Rubin. *Anti-American Terrorism and the Middle East*, Oxford, New York, 2002.

Sivan, Emmanuel. *Radical Islam: Medieval Theology and Modern Politics*, enlarged ed., Yale University Press, New Haven, 1990.

Weaver, Mary Anne. *Pakistan: In the Shadow of Jihad*, Farrar, Straus and Giroux, New York, 2002.

———. *A Portrait of Egypt*, Farrar, Straus and Giroux, New York, 1999.

Yousaf, Mohammad and Mark Adkin. *Afghanistan the Bear Trap: The Defeat of a Superpower*, Casemate Publishers, Havertown, Penn., 1992.

al-Zawahiri, Ayman. "Knights Under the Prophet's Banner," serialized in the London-based *Al Sharq Al Awsat* newspaper, December 2001.

Acknowledgments

Original research for this book was undertaken on four continents in more than twenty countries. I did much of this reporting from September 2001 through January 2003, while preparing a series of articles for the *Los Angeles Times*. I could never have produced those stories without the resources and support of a premier news-gathering organization such as the *Times*. For that, I am indebted to *Times* staff who graciously gave assistance when asked. I am further indebted to my editors for supporting and encouraging the reporting, even when it seemed—as it often did—fruitless. I would especially like to thank Dean Baquet for his encouragement and enthusiasm, Roger Smith for his patience, and John Carroll for his indulgence. The *Times* stories form the foundation upon which the book was written.

Additional invaluable reporting for those stories was accomplished by many *Los Angeles Times* colleagues. In particular, the book simply would not exist in anything like its present form without the reporting, insight, and relentless energy of Dirk Laabs, who began working on this project not long after I did and never stopped. In addition to unsurpassed reporting in Germany, he thought, fought, and suffered with me for every miserable fact and through every wretched turn of events for three years. Further indispensable and in many cases unmatchable reporting was accomplished by Patrick McDonnell in Kuwait and North Carolina; Josh Meyers in Pakistan and Washington, D.C.; Sebastian Rotella in Spain; Gil Reza in California; and Bob Drogin in Washington.

Additionally, I was assisted, and, as often, led by a global army of translators, friends, correspondents, drivers, and that peculiar agglomeration of all things helpful called fixers. Among them were Sol Vanzi,

Baradan Kuppusamy, Hany Fares, Jailan Zayan, Nagwa Hassaan, Paul Schemm, Ranwa Yehia, Leena Saeedi, Mohammed al-Asadi, Abdulhamid Alajami, Aly Rifaah, Ashraf Fouad Makkar, Cristina Mateo-Yanguas, Maribel Illescas, Yaqoub al-Mansour, Zulfiqar Ali, and Shamim ur-Rahman.

I thank my agent, Paul Bresnick, who, unless all agents provide their clients with ideas, titles, insight, and intellectual body armor, went far beyond the call of duty. I thank David Hirshey and Nick Trautwein at HarperCollins for the thoughtfulness, diligence, and obvious care with which they helped shape the book.

Above and beyond these people who helped in the actual construction of the book, I apologize to my family for my absences—at home and abroad—and thank them for their forbearance and especially for helping me see this through.

Index